Efficient Energy Utilization and Emission Reduction Strategies in Plant Operations

Efficient Energy Utilization and Emission Reduction Strategies in Plant Operations summarizes technological advancements in effective energy utilization in plant operations. It provides optimal geometrical parameters to design emission control equipment for minimum energy consumption.

Focusing on current and emerging technological innovations for improving the energy efficiencies of processes in several industries, such as manufacturing, mining, dairy, and other processing industries, this book explores pollution control and design methods for low-carbon environments. It also covers reductions in greenhouse gas emissions through innovative co-production of bio-oil in combined heat and power plants.

This book will be a useful reference for graduate students and academic researchers in the fields of energy management, plant operations, sustainable engineering, environmental engineering, and mechanical/industrial engineering.

Efficient Energy Utilization and Emission Reduction Strategies in Plant Operations

Edited by
S. Venkatesh, M.M. Matheswaran,
and S. Vijayan

CRC Press is an imprint of the
Taylor & Francis Group, an **informa** business

Designed cover image: Shutterstock

First edition published 2025
by CRC Press
2385 NW Executive Center Drive, Suite 320, Boca Raton FL 33431

and by CRC Press
4 Park Square, Milton Park, Abingdon, Oxon, OX14 4RN

CRC Press is an imprint of Taylor & Francis Group, LLC

ISBN: 978-1-032-58904-6 (hbk)
ISBN: 978-1-032-58905-3 (pbk)
ISBN: 978-1-003-45207-2 (ebk)

DOI: 10.1201/9781003452072

Typeset in Times
by codeMantra

Contents

Preface

Energy is a critical factor in production, necessitating management alongside traditional resources like land, labor, and capital. Energy-efficient production methods should be viewed as a practical and cost-effective means to maximize energy utilization given that saving energy is often significantly cheaper than generating it. Energy efficiency encompasses not only the physical performance of technical equipment and facilities but also the overall economic efficiency of the energy system, according to contemporary understanding.

In recent years, energy conservation and emission control have been heavily emphasized as government policies have become more stringent due to increased environmental concerns. Numerous technologies have been developed to improve energy utilization efficiency in various industries through technological innovations, and a lot of research efforts have been made to reduce the impact on the environment. It is feasible to achieve sustainable growth in major growing areas with limited energy supplies by using scientific and technological innovations to maximize energy potential. Academics are grappling with the challenge of connecting their critical technical innovations with emerging industries. This is necessary in order to meet the demand for increased energy efficiency and decreased emissions. A revolution in core technology is necessary if it is to reduce the amount of energy we waste while doing so. Thus, it is of utmost importance to improve the system for investing in technological innovation and to make it easier to establish fluid connections between research, technology, and money. Finally, sectors that are crucially important and that are still evolving should adopt multiple management modes for technological innovation, updating these modes regularly to align with the evolving technology paradigms.

The first section of this book examines technological innovations designed to improve energy efficiency across various industries, including oil and gas, digital manufacturing, food processing, and power plants. It begins with a thorough examination of energy utilization efficiency, emphasizing evaluations of utilities, industry, and transportation across different countries. Subsequent chapters focus on advancements within the oil and gas sectors, coal-fired power stations, digital twin technologies in manufacturing, and energy utilization in food processing. Furthermore, these chapters offer a comprehensive analysis of significant technological advancements that are essential for establishing a low-carbon world.

The second section concentrates on cutting-edge emission control techniques, such as computational fluid dynamics for pollution control components and geographic information systems studies of pollutants in specific regions of India. This section also discusses the role of smart grid technologies in waste reduction plant operations and advanced emission control technologies in foundries.

Finally, the third section of the book addresses the broader policy framework that supports technological innovation for sustainable development. It provides a global perspective on innovation support policies, examining the historical context, the evolution of research and innovation policies, and their alignment with sustainable development goals.

This comprehensive collection intends to be a useful resource for industry professionals, scholars, and policymakers. We hope that the ideas offered in this book will contribute to the continuing discussion about energy efficiency, technological innovation, and environmental stewardship which will direct future progress in all the above crucial areas.

About the Editors

Dr. S. Venkatesh is a Professor in the Department of Mechanical Engineering, Sri Eshwar College of Engineering, Coimbatore, Tamil Nadu, India. He completed his post-doctorate in the area of multi-cyclone arrangement for emission control at the National Institute of Technology, Tiruchirappalli, Tamil Nadu, India, in 2022 as institute-funded category. He completed his doctor of philosophy in the area of emission control equipment called cyclone separator from Anna University, Chennai, India, in 2018. He has completed his bachelor's degree in Mechanical Engineering in 2006. He has completed his master's degree in Engineering Design in 2013. He received a grant for Rs. 22.5 lakhs from the National Institute of Technology, Tiruchirappalli, Tamil Nadu, India, for pursuing his post-doctorate. In 2024, he received an 18.3-lakh grant from the SERB-TARE scheme sponsored by the Government of India. He is a Co-PI on the DST FIST project (a 59-lakh grant was received from DST, Government of India). He has 12 years of teaching experience, 3 years of industry experience, and 2.6 years of postdoctoral work experience. He is a recognized research supervisor at Anna University Chennai. Four research scholars are pursuing Ph.D. under his guidance. He is a researcher in emission control equipment called gas cyclone separator, hydro-cyclone, and Venturi scrubber. Moreover, he is a researcher in particle flow analysis, CFD, multiphase flow, and optimization of industrial emission control equipment geometries. His research work is applicable to control pollution in various industries such as chemical engineering industries, foundry, cement factory, and coal mining sector. He is also a recognized reviewer in many reputed SCI journals. He got an outstanding reviewer award for reviewing the manuscript from the *Journal of Aerosol Science* in 2018. He got the best researcher award from Sri Eshwar College of Engineering, Coimbatore, Tamil Nadu, India, in 2018. He got the top 5% researcher award in 2024 from Sri Eshwar College of Engineering, Coimbatore. He has published many research articles in SCI and Scopus-indexed journals. He has published a book titled *The Gas Cyclone Separator: A Review* in 2020. Moreover, he has published a book titled *An Introduction to Safety and Risk in Industries* in 2021. He has published book chapters in reputed publications. He edited a special edition titled *Innovations in Energy Utilization and Equipment Design* in 2024 which was published in the *Pertanika Journal of Science and Technology* indexed in Scopus. He is an editor for the First International Conference on Emerging Trends in Mechanical Sciences for Sustainable Technologies-2023. He is a convenor and editor for the First International Conference on Sustainable Materials and Technologies-2024. He is the editor-in-chief of the *Journal of Advanced Mechanical Sciences.*

Dr. M.M. Matheswaran is an accomplished academic and researcher, currently serving as an Associate Professor in the Department of Mechanical Engineering at Jansons Institute of Technology, Coimbatore. He has a strong academic background, having completed all his degrees at Anna University, a prestigious institution in Tamil Nadu, India. He earned his Bachelor of Engineering (B.E.) degree in 2006, followed by a Master of Engineering (M.E.) in 2009, both specializing in Mechanical Engineering. He further advanced his expertise by obtaining a doctor of philosophy (Ph.D.) in 2020, with his research focusing on solar thermal systems. With over 15 years of diverse experience spanning industry, teaching, and research, he has significantly contributed to the field of mechanical engineering, particularly in the areas of solar thermal energy and thermal engineering. His research has led to the publication of 25 papers in esteemed peer-reviewed international journals, conferences, and book chapters. His scholarly work is highly regarded in the academic community, demonstrating his deep commitment to advancing knowledge in his areas of expertise. He is also recognized for his editorial contributions. He edited the book *Solar Thermal Conversion Technologies for Industrial Process Heating*, published by CRC Press, which is a significant resource in the field of renewable energy. He is currently working on another editorial project, *Energy Utilization and Emission Control Techniques in Plant Operations*, for CRC Press, further underscoring his leadership in the field. In addition to his academic achievements, he has demonstrated a strong inclination toward innovation. He has filed six patents, with three at the international level and three at the national level, showcasing his ability to translate research into practical applications. His innovations primarily focus on solar thermal technology, reflecting his dedication to advancing sustainable energy solutions. His research interests are centered on solar thermal conversion systems, thermal energy storage for low- and medium-temperature applications, and heat transfer. His work in these areas has not only advanced academic understanding but also led to tangible outcomes. He has been awarded grants totaling Rs 17,00,000 through prestigious competitions such as the Onion Grand Challenge, Idea for Change Challenge, and Yukti Innovation Challenge, all aimed at product development in the field of solar thermal technology. His editorial work also extends to academic journals, where he recently edited a special issue for the *Pertanika Journal of Science and Technology* (PJST) (Special Issue – IEUED 2023), Vol. 32 (S1) 2024. This role highlights his standing in the academic community as a thought leader in his field. Overall, he is a distinguished figure in the field of mechanical engineering, particularly in the domains of solar thermal energy and thermal engineering. His work continues to have a significant impact on both academia and industry, and his contributions to research, innovation, and education are widely recognized.

Dr. S. Vijayan is currently working as an Associate Professor in the Department of Mechanical Engineering at the Coimbatore Institute of Engineering and Technology, Coimbatore, India. He received his bachelor's, master's, and Ph.D. degrees from Anna University in 2006, 2011, and 2018, respectively. Driven by a deep passion for research, he pursued postdoctoral studies at the National Institute of Technology, Tiruchirappalli, specializing in solar thermal energy storage, from December 2019 to May 2022. With 15 years of experience in industry, academia, and research, he has made significant contributions to the fields of energy utilization and emission control. His research has been pivotal in advancing technologies that improve energy efficiency and reduce environmental impact. As a co-principal investigator, he completed a Department of Science and Technology (DST), India-funded research project titled "Development of Solar Evacuated Tube Collectors with Inserted Baffles for Direct Air Heating in Industrial Applications," with a total project cost of Rs. 1.62 crores. Additionally, he was awarded a SERB project under SURE, securing a grant of Rs. 23.76 lakhs over 3 years as the co-principal investigator. He has authored more than 32 peer-reviewed international journal publications, 15 international and national conference papers, and eight book chapters with prominent publishers such as Springer Nature, CRC Press, and Elsevier, among others. His editorial work includes the book *Solar Thermal Conversion Technologies for Industrial Process Heating* for CRC Press, and he has also edited special issues for Scopus-indexed journals. He holds six granted patents, including three international and three national patents. In addition to his research and academic endeavors, he serves as an associate editor for the *Journal of Advanced Mechanical Sciences* (JAMS) and as a reviewer for various reputed international journals. His current research interests focus on solar thermal conversion systems, thermal energy storage for low- and medium-temperature applications, and advanced heat transfer techniques.

Contributors

T.V. Arjunan
Department of Mechanical Engineering
Guru Ghasidas Vishwavidyalaya
Bilaspur, Chhattisgarh, India

M. Arulraj
Department of Mechanical Engineering
Sri Krishna Polytechnic College
Coimbatore, Tamil Nadu, India

P. Arunachalam
Department of Mechanical Engineering
Government Polytechnic College
Kancharapalem, Visakhapatnam, India

Manjushree Banerjee
Independent Researcher
New Delhi, India

S. Ganeshkumar
Department of Mechanical Engineering
Sri Eshwar College of Engineering
Coimbatore, Tamil Nadu, India

G. Gokilakrishnan
Centre for Advanced Material Processing and Testing
Sri Eshwar College of Engineering
Coimbatore, Tamil Nadu, India

K. Harisivasri Phanindra
Department of Mechanical Engineering
Vel Tech Rangarajan Dr. Sagunthala R&D Institute of Science and Technology
Chennai, Tamil Nadu, India

Sivaramasundaram Krishnamoorthy
Department of Physics
Sethu Institute of Technology
Virudhunagar, Tamil Nadu, India

Prashant Kumar Jangde
Department of Mechanical Engineering
Guru Ghasidas Vishwavidyalaya
Bilaspur, Chhattisgarh, India

Shalini Kumari
Department of Management,
Birla Institute of Technology, Mesra
Lalpur, Ranchi, India

R. Manikandan
Department of Mechatronics Engineering
Sri Krishna College of Engineering and Technology
Coimbatore, Tamil Nadu, India

M.M. Matheswaran
Department of Mechanical Engineering
Jansons Institute of Technology
Karumathampatti, Tamil Nadu, India

Nagaprasad Nagaraj
Department of Mechanical Engineering
ULTRA College of Engineering and Technology
Madurai, Tamil Nadu, India

Bhaskar Natarajan
Energy Efficiency Specialist
New Delhi, India

Pradeep Patanwar
Department of Mechanical Engineering
Guru Ghasidas Vishwavidyalaya
Bilaspur, Chhattisgarh, India

R. Prakash
Department of Mechanical Engineering
Velalar College of Engineering and Technology
Erode, Tamil Nadu, India

L. Priyanka Dwarampudi
Department of Pharmacognosy
JSS College of Pharmacy, JSS Academy of Higher Education & Research
Mysuru, Karnataka, India

V. Rajkumar
Department of Mechanical Engineering
Coimbatore Institute of Engineering and Technology
Coimbatore, Tamil Nadu, India

T. Ramakrishnan
Centre for Advanced Material Processing and Testing
Sri Eshwar College of Engineering
Coimbatore, Tamil Nadu, India

Krishnaraj Ramaswamy
Department of Mechanical Engineering
Dambi Dollo University
Ethiopia

Shanmugam Ramaswamy
TIFAC, CORE-HD, Department of Pharmacognosy
JSS College of Pharmacy, JSS Academy of Higher Education & Research
Mysuru, Karnataka, India

R. Rameshbabu
Department of Mechanical Engineering
Karpagam Institute of Technology
Seerapalayam, Tamil Nadu, India

Vicky Saurabh
Department of Operations Management
Indian Institute of Management,
Ranchi Jharkhand, India

S.P. Sivapirakasam
Department of Mechanical Engineering
National Institute of Technology
Tiruchirappalli, Tamil Nadu, India

M. Sowrirajan
Department of Mechanical Engineering
Coimbatore Institute of Engineering and Technology
Coimbatore, Tamil Nadu, India

R. Suresh Kumar
Centre for Advanced Material Processing and Testing
Sri Eshwar College of Engineering
Coimbatore, Tamil Nadu, India

V. Sureshkannan
Department of Mechanical Engineering
Builders Engineering College
Tirupur, Tamil Nadu, India

Mostafa Tahmasebi
Department of Mechanical Engineering
University of Cincinnati
Cincinnati, OH, USA

L. Venkatesh
Department of Mechanical Engineering
Coimbatore Institute of Engineering and Technology
Coimbatore, Tamil Nadu, India

S. Venkatesh
Centre for Advanced Material Processing and Testing
Sri Eshwar College of Engineering
Coimbatore, Tamil Nadu, India

S. Vijayan
Department of Mechanical Engineering
Coimbatore Institute of Engineering and Technology
Coimbatore, Tamil Nadu, India

A. Vimal
Department of Mechanical Engineering
Sri Eshwar College of Engineering
Coimbatore, Tamil Nadu, India

R. Vivek
Department of Mechanical Engineering
Sri Eshwar College of Engineering
Coimbatore, Tamil Nadu, India

1 Overview of Country-Wise Energy Utilization Efficiency

Mostafa Tahmasebi

1.1 INTRODUCTION

The world's energy demand is increasing due to population growth and the spread of industrialization across the globe. However, this growth comes with a significant challenge: how to meet this demand while avoiding harm to our environment or depleting our resources. The answer may perhaps be in improving how efficiently we use energy, especially in industries that are major energy consumers.

Recent data from the International Energy Agency (IEA) indicates that worldwide energy usage surged by over 2.3% in a single year, marking the fastest pace of growth in the last decade [1]. This demand is expected to increase faster in the near future going from 2.2% in 2023 to 3.4% by 2026 [2]. This demand is driven largely by economic expansion and increased heating and cooling needs. While this increase in energy use indicates progress and development, it also raises concerns about the sustainability of our energy practices and the health of our planet.

There's been a noticeable shift toward renewable energy sources in the last few years, which is a positive step toward reducing our environmental footprint. However, the transition to greener energy isn't happening quickly enough to offset the rising demand as renewables currently meet only about 25% of the increase in global energy demand [3]. This is where energy efficiency comes into play. The IEA suggests that with robust energy efficiency policies, we could achieve over 40% of the emission reductions required to align with the Paris Agreement goals, without relying on new technology [4].

However, there's a significant difference in energy consumption patterns between developed and developing countries. Developed countries, despite having a smaller portion of the world's population, use a much larger share of its energy. This imbalance highlights the need for more efficient energy use worldwide, not just in countries with high consumption rates.

Developing countries are rapidly increasing their energy use as their economies grow. Countries like China and India, for example, have seen their energy demand grow substantially, contributing significantly to the global increase. This increase is essential for improving living standards but comes with the challenge of relying heavily on fossil fuels, which contribute to pollution and climate change. Transitioning to more sustainable energy sources and improving energy efficiency

DOI: 10.1201/9781003452072-1

are crucial for these countries to meet their development goals without compromising the environment.

As this chapter begins to explore energy utilization efficiency, it will examine the overarching view of worldwide energy consumption and delve into particular sectors that significantly contribute to energy demand. This chapter will explore the latest technological and policy advancements aimed at enhancing energy efficiency and minimizing environmental impact. By understanding the current state of energy use and the potential for efficiency improvements, the significance of this matter and its influence in guaranteeing a sustainable future for our world will be more fully appreciated [5].

1.2 GLOBAL ENERGY LANDSCAPE

In recent years, the world's energy consumption has increased significantly. For example, in 2022, we used about 460 terawatt-hours (TWh) according to the IEA [2]. This significant figure shows just how much energy is needed for everything we do, from keeping our homes warm to running factories.

Looking at how this energy is utilized, the difference between different parts of the world is clear. Developed areas such as North America and Europe, with a relatively small percentage of the global population, account for around 20%, a significant chunk of energy use. This high level of consumption is linked to the advanced industrial activities and higher living standards in these regions.

On the other hand, in developing regions like Asia, Africa, and South America, energy consumption is rapidly increasing. In particular, Asia's share of the world's energy use has gone up to more than 40%, with countries like China and India using more and more energy as they grow and improve the lives of their people [3]. As depicted in Figure 1.1, a historic shift will occur by 2025, where Asia will be responsible for half of the world's electricity usage. Additionally, one-third of the worldwide electricity consumption will be attributed to China.

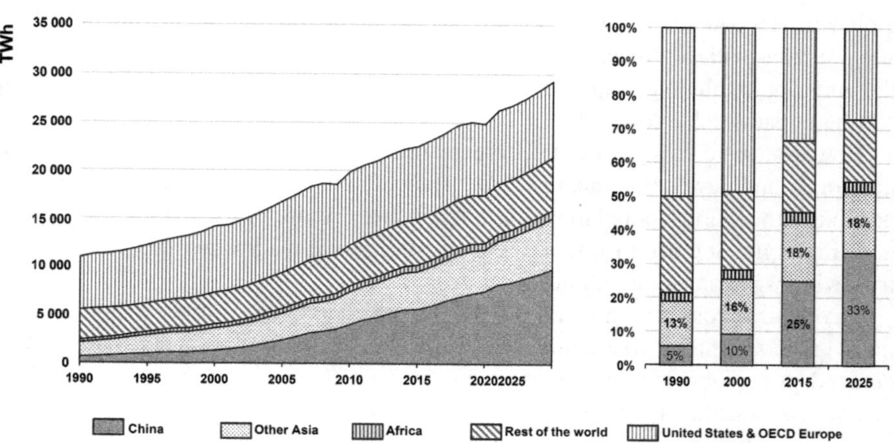

FIGURE 1.1 Evolution of global electricity demand by region and regional shares 1990–2025. (According to IEA Electricity Market Report 2023.)

Looking at how energy is used, the industrial and transportation sectors are the primary consumers, with about 35% of the total for the industrial sector and 36% for the transportation sector. Homes and businesses together use about 22%, which covers energy needed for daily life, from heating homes to powering businesses [3]. Despite initiatives to improve efficiency, the demand for energy in the industrial sector continues to rise. This is especially true in developing nations where the rate of industrialization is growing. For the transportation sector, the transition to electric vehicles (EVs) and alternative energy sources such as biofuels and hydrogen is initiating a transformative phase. This change is potentially viable to minimize carbon emissions and improve energy efficiency, and both are beneficial. In the residential and commercial sectors, enhanced building designs, energy-saving lighting and HVAC systems, and the incorporation of intelligent technologies for energy management can lead to improved energy efficiency. The push for green buildings and sustainable architecture is aiding in reducing the energy demand of commercial spaces.

1.3 ENERGY PRODUCTION AND SOURCES

The worldwide energy industry continues to depend largely on fossil fuels, contributing to approximately 80% of our overall energy usage. Despite their historical role in driving economic growth, the environmental impact of coal, oil, and natural gas, particularly their carbon emissions, is pushing the industry toward renewable alternatives [6]. Fossil fuels continue to dominate global energy supplies due to their established infrastructure and reliable output. Regions that are relying heavily on fossil fuel industries face additional socio-economic challenges. This requires us to come up with effective solutions to manage this shift, ensuring we balance economic and environmental needs.

The landscape of renewable energy is experiencing rapid growth, fueled by technological advancements and cost reductions, particularly in solar and wind energy [7]. The decline in solar panel costs by over 80% since 2010 is an example of this trend. By 2023, renewables were contributing around 30% to global electricity generation, showing how fast this sector is growing [8]. This growth in renewables highlights the sector's potential and competitiveness against traditional energy sources. However, moving from fossil fuels to renewables isn't straightforward, particularly in unpredictable energy sources like solar and wind. This variability calls for the development of robust energy storage solutions and more flexible grid systems. Alongside these challenges, significant investments are required to build out renewable energy infrastructure. Particular to regions dependent on fossil fuel, there's the added challenge of finding new jobs and opportunities for those affected by the shift.

Looking to the future, we're likely to see a more varied mix of energy sources, prioritizing sustainability, efficiency, and security. Innovations in energy storage, such as battery technologies, grid management, and clean energy technologies, alongside supportive policies, will be crucial for accommodating the variability of renewable sources.

1.4 ENERGY DEPENDENCY AND SECURITY

Energy dependency varies significantly across the world, with some regions relying heavily on imports to meet their energy demands. As an example, the European Union (EU) purchases more than 50% of its energy from other countries.

Some countries within the EU depend even more heavily on these imports, with figures reaching up to 80%. This high level of dependency indicates that this region can be vulnerable to external market fluctuations and geopolitical tensions which became evident through the Russia-Ukraine war. In another example, countries heavily reliant on imported oil and gas, such as Japan and South Korea, face the risk of price volatility, which can have ripple effects on their economies. In 2020, global oil price fluctuations demonstrated how dependent economies could be affected by shifts in the energy market [9,10].

To avoid these risks, countries are actively seeking to diversify their energy sources. The United States is a good example of this shift by reducing its dependence on foreign oil, with domestic production meeting nearly 80% of its needs in recent years [11]. This achievement was facilitated by progress in oil and gas extraction methods, coupled with transformative investments in energy efficiency research and the implementation of renewable energy sources. The United States' Strategic Petroleum Reserve can hold up to 727 million barrels of oil, which demonstrates how diversifying energy sources enhance energy security for nations [12,13]. This requires a collective effort by countries to work individually and together to shape energy policies that encourage energy efficiency, endorse the adoption of renewable energy, and promote research and development in new energy technologies that can ultimately benefit all and contribute to greater energy independence.

1.4.1 Economic and Political Influences on Energy

Economic expansion is a key catalyst for energy demand. As economies flourish, their requirement for energy to power industrial activities, transport, and household consumption escalates. Rapidly growing economies, especially in Asia, have seen significant rises in energy consumption. For instance, over the past several decades, China and India have seen significant economic expansion, positioning them as central figures in the international energy markets [14].

Energy markets are highly sensitive to supply and demand dynamics, geopolitical events, and regulatory changes. Significant fluctuations in oil prices, such as the 2014 oil price crash and subsequent recoveries, illustrate how excess supply or geopolitical tensions can impact global energy prices, affecting economies worldwide. Government policies and political decisions play crucial roles in shaping the energy sector. Initiatives to reduce carbon emissions, subsidies for renewable energy, and regulations on fossil fuel extraction and use significantly influence the energy mix and market trends [15]. For example, the Paris Agreement, signed and endorsed by more than 190 countries, seeks to control global warming, spurring nations to set more ambitious targets for renewable energy adoption [16].

Economic incentives and political will are driving the transition toward renewable energy. Investments in renewable energy technologies are increasing. In the international arena, the capacity additions for renewable electricity hit an estimated 507 gigawatts (GW) in 2023, nearly 50% more than in 2022. This surge was propelled by sustained policy backing in over 130 countries, sparking a notable shift in the global growth trajectory [17]. Some countries are leading this effort; for example, in 2020 alone, China invested approximately $83.4 billion in renewable energy, accounting for about 30% of the global total investment in renewable energy sources. This investment has been directed mainly toward solar and wind power projects, solidifying China's position as a global leader in renewable energy capacity [18]. This transition is made possible by both the decreasing cost of renewables and government policies supporting clean energy development. This shift is reshaping the energy landscape, with implications for traditional energy powerhouses and emerging renewable leaders.

1.4.2 Challenges and Opportunities

As the chapter examines the energy sector today, it highlights a mix of challenges and exciting possibilities. The push for a cleaner, more efficient way of using energy is ongoing, but it is not without its hurdles.

One of the biggest challenges is cutting down the carbon emissions from energy production, especially as the world's demand for energy keeps growing. This is even more pressing for fast-developing economies that need more energy to fuel their growth. On the bright side, there are plenty of opportunities too, especially when it comes to energy storage. Experts believe we could see up to $620 billion invested in energy storage by 2040. This is crucial for making renewable energy more dependable and keeping our power grids stable [19].

1.4.3 Emergence of Energy Efficiency in Different Countries

Having reviewed energy in a global landscape, this section will take a closer look at how individual countries are charting their unique paths toward a sustainable energy future. Across the world, nations are embracing the benefits of energy efficiency with a blend of innovation, policy, and strategic investments. Each country below is a pioneer in terms of their approach to energy efficiency and or is a significant energy consumer, contributing to a large fraction of energy use worldwide. Their methods are shaped by their unique set of resources, economic structures, and environmental commitments, leading to a diverse array of strategies and outcomes.

1.4.3.1 China

In China, even though the need for electricity is rising, there was a significant decline in the power sector's emission intensity in 2022, dropping by 2.5% [20]. China's ambitious commitment to achieve carbon neutrality by 2060, together with its goal to quadruple its economy's size by 2035, sets a strong foundation for its energy transformation. This transition is backed by the latest Five-Year Plan, targeting an 18% cut in carbon dioxide (CO_2) intensity and a 13.5% reduction in energy intensity from 2021 to 2025 [21]. This presents substantial prospects for sectors that align with China's industrial and environmental goals [22]. Despite the decline in coal's share of energy consumption from 68.5% to 56% over the last decade, recent challenges such as power rationing and the impacts of droughts on hydropower production have necessitated a more cautious approach to reducing coal dependency [23].

In 2021, petroleum and other liquid fuels remained the second major source of energy in China, making up 19% of the nation's total energy consumption. While China has made significant progress in diversifying its energy composition and embracing cleaner fuels, hydroelectric sources, natural gas, nuclear power, and non-hydro renewables still constitute minor segments of its energy portfolio [23]. However, the usage of natural gas, nuclear power, and renewable energy in China is on a rising trend, aiding in the gradual transformation of the country's energy scenario.

Natural gas, in particular, has seen remarkable growth, becoming China's fastest-growing primary fuel with demand quadrupling over the past decade. As the globe's third-largest consumer of natural gas and the top importer in 2021, China's natural gas sector plays a crucial role in diminishing dependency on coal. The efforts to raise the contribution of natural gas in the energy blend to 15% by 2030 will require significant imports, both via pipelines and in the form of liquefied natural gas (LNG), to fill the disparity between domestic output and consumption.

The focus on carbon capture, utilization, and storage (CCUS) technology highlights China's commitment to innovative solutions for achieving its carbon neutrality goals. With the potential market for CCUS expected to exceed 100 billion yuan by 2030 and reach 330 billion yuan by 2050, China is positioned to become a significant player in the global CCUS landscape. This technology, particularly direct air capture, offers promising avenues for reducing atmospheric CO_2 levels and is anticipated to serve a pivotal part in China's decarbonization efforts [24].

China's average annual investment of USD 75 billion in its grid since 2010 has significantly improved the integration of renewable energy sources, reducing the curtailment of wind and solar power. The emphasis on solar photovoltaic (PV), which accounted for 70% of renewable capacity additions in 2022, along with investments in wind and hydropower, highlights China's strategic approach to expanding its renewable energy infrastructure. This commitment to renewable energy development, coupled with efforts to diversify energy imports and advance CCUS technologies, illustrates China's multifaceted strategy for navigating its energy transition and achieving its ambitious environmental and economic goals [24,25].

1.4.3.2 India

India, as the third-largest energy consumer globally, is at a pivotal point in its energy journey, with a record demand of 223 GW in 2023, marking a 3.4% increase from the previous year. The country's energy landscape is diverse, consuming a mix of traditional sources like coal, oil, and gas, alongside renewable and sustainable options such as solar, wind, biomass, and hydro plants. With a booming economy and a population of approximately 1.4 billion, India's rapid urbanization and industrial growth are driving an ever-increasing demand for energy [26].

Despite the dominance of fossil fuels in India's power sector, the nation has set forth ambitious objectives to elevate the role of renewable and nuclear energy significantly. The government's pledge, conveyed through the "Panchamrit" strategy announced at the Conference of the Parties (COP) 26 climate conference, involves attaining 500 GW of non-fossil energy capacity and meeting 50% of its energy needs with renewables by 2030. These targets highlight India's dedication to mitigating greenhouse gas emissions and reducing dependency on energy imports, thereby increasing national energy security. However, the transition from coal to cleaner energy sources is anticipated to span decades, necessitating a balanced approach that leverages fossil fuels, particularly natural gas, to ensure grid stability [27].

India's renewable energy capacity stood at 179 GW as of May 31, 2023, with solar and wind energy making up significant portions. The disparity between installed renewable capacity and actual electricity production from non-fossil sources highlights the challenges posed by the variability of renewable energy. India's strategy to increase the portion of natural gas in its energy portfolio to 15% by 2030 and its status as a major importer of LNG and crude oil further illustrate the complexities of transitioning to a more sustainable energy portfolio [28].

The government's "Make in India" program, designed to boost local manufacturing, presents both challenges and opportunities for foreign companies, including those from the United States. Despite preferences for Indian suppliers in government tenders, the United States has become a crucial energy ally for India, with the exchange of energy commodities and equipment hitting $18.5 billion in 2022. This growth is propelled by India's surging energy demand and collaborative commitments to clean energy [29].

In the transmission, distribution, and smart grid sector, India faces challenges such as high technical and commercial losses and the financial distress of distribution companies. To address these issues, the government has launched reform-based programs and aims to install 250 million smart meters by 2025, enhancing grid efficiency and utility revenue [30]. The renewable energy and energy storage sectors are ready for rapid expansion, supported by government tariffs and incentives that encourage domestic manufacturing and reduce import reliance. India's ambition to manufacture solar PV cells and become a hub for wind energy components spotlights its potential as a significant player in the global renewable market. India's LNG imports and the government's vision for a "One Nation, One Gas Grid" show the importance of natural gas in the country's energy transition. The expansion of gas pipeline infrastructure and city gas distribution projects offer substantial opportunities for collaboration and investment, particularly from U.S. firms. Moreover, India's commitment to the green hydrogen economy and the National Hydrogen Mission highlights the country's proactive approach to embracing emerging clean energy technologies.

1.4.3.3 Germany

Germany's Energiewende, a policy framework initiated to transition toward a renewable and efficient energy system, places a strong emphasis on wind and solar power. This ambitious program aims not only to reduce the dependence on fossil fuels but also to cut carbon emissions significantly. In 2021, Germany allocated 28 billion euros toward investments in renewable energy, predominantly focusing on the accelerated deployment of wind and solar power. The nation has set a target to generate 80% of its electricity from renewable sources by 2030 and aspires to meet its entire electricity demand through renewable sources by 2035 [31]. Despite there being a slight 2% decrease in greenhouse gas emissions across all sectors in 2022, this reduction is insufficient in achieving the ambitious 2030 climate goal of cutting emissions by 55% compared to the levels recorded in 1990 [32]. The energy sector's response to geopolitical tensions, notably Russia's conflict with Ukraine, led to an increased reliance on coal, complicating efforts to decrease emissions.

In pursuit of its energy targets, Germany is advantageously placed to actualize these objectives by channeling significant investments into offshore wind and PV projects, expanding grid infrastructure, and improving energy storage solutions. The country's long-standing leadership in renewable energy and environmental technologies provides a solid foundation for these future endeavors [33]. Despite these advancements, Germany's energy security is challenged by its dependence on imported oil and gas, which accounted for nearly all its consumption, making the nation susceptible to global market fluctuations and geopolitical dynamics.

During 2022, Germany's primary energy usage reached 11,769 petajoules, with a considerable 75% deriving from fossil-based sources. The energy mix saw mineral oil at 35.3% and natural gas at 23.7%, while renewables made up 17.2% [34]. The push for renewable energy is evident in the electricity sector, where 48.3% of the electricity came from renewable sources in 2022, a 5.6% increase from the previous year. Wind power led the renewables with a 25.9% share, producing about 131.3 TWh [35].

The country's energy policy, steered by the Federal Ministry for Economic Affairs and Climate Action, is focused on expanding renewable energy and enhancing efficiency [36]. Germany's approach includes significant legal frameworks like the Renewable Energy Source Act (EEG) to promote renewable electricity [37]. The energy landscape is also shaped by Germany's imports, with natural gas imports totaling 1,449 TWh in 2022, showing a diversification in supply sources in response to reduced deliveries from Russia [38].

Opportunities in the energy sector are vast, particularly in energy storage and smart grid technologies, essential for integrating fluctuating renewable sources. The country's energy infrastructure investments and the pioneering Hydrogen Strategy highlight the proactive steps being taken to ensure a sustainable and secure energy future. By 2030, Germany expects its domestic hydrogen demand to fall within the 90–110 TWh range, with forecasts suggesting that approximately two-thirds of this usage will be satisfied through imports [39]. The strategic focus on hydrogen imports, primarily through pipelines from European neighbors and overseas shipments from regions like the United States and Persian Gulf countries, highlights the outlook of Germany's hydrogen ambitions. In 2022, Germany imported roughly 1.32 million tons of LNG, and it is anticipated that a substantial part of these imports will shift toward hydrogen in the future [40].

1.4.3.4 Norway

Norway, along with its Nordic neighbors, stands at the forefront of renewable energy utilization, green technology adoption, and sustainable resource management practices [41]. Despite its significant role as a major oil and gas exporter, Norway faces challenges in curbing CO_2 emissions due to the carbon-intensive nature of oil and gas production. The majority of Norwegian society is already powered by clean hydropower, leaving limited scope for drastic emission reductions. The transportation sector, however, has seen Norway take significant strides in electrification.

Over the past century, Norway's extensive hydro resources have been harnessed through the construction of more than 330 dams. This foundation of zero-emission electric power creates favorable conditions for achieving carbon emission neutrality across the entire value chain, encompassing hydrogen, ammonia, battery production, metal production, and various transportation methods. Nevertheless, these hydro systems require modernization to continue their efficiency [42].

The hydrogen and ammonia sectors are witnessing rapid expansion, with an increasing number of companies venturing into the hydrogen field. Recognized by the government as a crucial component of the green transition, hydrogen has received significant R&D funding and support. Similarly, ammonia is being considered as a potential clean fuel for maritime shipping.

In the battery sector, Norway is planning the establishment of gigafactories, aiming to source materials like cobalt and nickel locally or innovate in battery composition to lessen the reliance on imported rare earth minerals. With a high domestic demand for batteries, Norway and its regional partners are looking to decrease dependency on external sources in this critical value chain.

Regarding wind energy, Norway has ambitious plans to develop 30 GW of offshore wind capacity by 2040, which would make it a leader in renewable energy production per capita. Despite facing opposition to land-based wind projects, a new tax regime benefiting local communities might facilitate the permitting process [43].

Shifting the focus toward EVs, Norway's substantial incentives have led to electric drivetrains dominating about 90% of new passenger car sales in the consumer market. The government's new vehicle procurement regulations and high ambitions for electric ferries and aviation further underscore Norway's commitment to electrification across various transport sectors [44].

Norway is also pioneering in next-generation agriculture, focusing on sustainable technologies and innovative farming methods to reduce emissions, maintain soil health, and protect biodiversity. Supported by governmental initiatives and cross-border collaborations, these new business models are adapting to Norway's high labor costs through increased automation.

Carbon capture and storage (CCS) is a priority for Norway, which has been instrumental in developing CCUS technologies through collaborations like those with the U.S. Department of Energy at Test Center Mongstad (TSM). Norway is scaling these concepts globally, with projects like the Heidelberg cement factory at Heroya being among the initial capture sites. The "Longship" project, an undersea carbon storage initiative, is set to launch in 2024 with a capacity to store up to 1.5 million tons of CO_2 annually. In sustainable fuels, Norway's vast forest resources, which have been underutilized for fuel production, present opportunities for developing advanced biofuels [45]. The airport operator Avinor suggests that forest by-products could significantly contribute to the fuel needs of Norwegian aviation. Research into sustainable aviation fuels (SAF) and biomass sources like seaweed and algae for various applications is driving the need for international collaboration in sustainable fuel development.

The emphasis on green buildings and intelligent infrastructure within Norway's smart city initiatives highlights the prioritization of energy efficiency and artificial intelligence (AI) integration in creating sustainable urban environments. The focus extends to innovative public transport systems designed to integrate seamlessly with existing networks. While Norway seeks international partnerships for technology adoption, competition, particularly from Asia, remains significant.

1.4.3.5 United States of America

The United States is undertaking significant strides under the Biden administration to lead the global charge in addressing the climate crisis, with a comprehensive strategy that intertwines federal procurement power and operational transformation. This initiative is not merely about reducing emissions within federal domains but also about catalyzing the growth of the clean energy sector, fostering resilient communities, and ensuring economic competitiveness through sustainable job creation and industry growth.

At the heart of this initiative are ambitious targets that reflect a deep commitment to environmental stewardship and sustainable development [46]:

Transitioning to 100% carbon pollution–free electricity (CFE) by 2030 is a bold step, aiming to significantly reduce the carbon footprint of federal operations and encourage the uptake of clean energy across the nation [47].

The move toward 100% zero-emission vehicle (ZEV) acquisitions by 2035, with an interim goal for light-duty vehicles by 2027, aligns with broader efforts to electrify transportation and reduce reliance on fossil fuels [48].

Achieving net-zero emissions in federal procurement by 2050, particularly through the use of low-emission construction materials, signifies a dedication to practices that promote sustainable development.

The goal of a net-zero emissions building portfolio by 2045, with a significant reduction in emissions by 2032, sets new standards for energy efficiency and sustainable architecture [49].

Aiming for net-zero emissions from all federal operations by 2050, with a substantial reduction target by 2030, highlights the administration's holistic approach to carbon neutrality [50].

These targets are part of a larger vision that encompasses climate-resilient infrastructure, a workforce focused on sustainability, environmental justice, and the prioritization of sustainable products. The administration's approach leverages domestic and international partnerships to amplify its impact, demonstrating a commitment to collaborative action in tackling global environmental challenges.

In support of these goals, several federal agencies have launched clean energy projects and initiatives that exemplify the United States' dedication to cleaner, more affordable energy. For instance, the record-breaking offshore wind lease sale conducted by the Department of the Interior in the New York Bight signifies a major leap in clean energy generation, with the potential to power millions of homes and create thousands of jobs. This initiative showcases the Biden administration's coordinated effort to address climate change, promote economic growth through jobs, and accelerate the shift to a clean energy economy. The United States' efforts are not limited to domestic policy changes but extend to international collaborations and commitments, reinforcing its standing as a frontrunner in the worldwide shift toward sustainability and climate resilience. Through these all-encompassing strategies and initiatives, the United States is striving to construct a sustainable future. This future not only responds to the immediate climate change challenges but also sets the foundation for enduring environmental and economic prosperity.

1.4.4 Policy Framework and International Cooperation

Across the globe, energy policies have been instrumental in driving significant shifts:

The addition of renewable energy capacity worldwide has experienced unparalleled growth, with the International Renewable Energy Agency (IRENA) documenting an extraordinary increase of over 260 GW in the year 2020 alone [51]. This surge is primarily attributed to the global commitment toward transitions to cleaner energy sources [52].

Advancements in energy efficiency have led to marked declines in energy intensity within major global economies. The IEA notes that between 2000 and 2020, improvements in energy efficiency contributed to a global reduction in energy intensity by 15%, emphasizing the impactful nature of such efficiency enhancements and technological innovations.

By the year 2050, the EU intends to achieve carbon neutrality through the implementation of its ambitious Green Deal. This initiative has already yielded a significant decrease in greenhouse gas emissions, achieving a 24% reduction from the levels recorded in 1990 by the year 2020, demonstrating the potency of unified regional policy initiatives [53,54].

The worldwide average surface temperature is 1.2°C greater than pre-industrial times. This increase has led to a higher frequency of heatwaves and other extreme weather events [55]. The energy sector remains the largest contributor to air pollution, which affects over 90% of the global population and is associated with more than 6 million premature deaths each year. Despite efforts to curb emissions, greenhouse gas concentrations have not yet peaked, highlighting the critical need for effective measures to address climate change and reduce air pollution. Recent trends indicate a slowdown or even a reversal effort being made in expanding access to clean cooking and electricity solutions in certain regions [56].

In this context, the emergence of a new clean energy economy, primarily driven by solar PVs and EVs, presents a glimmer of hope. Since 2020, investment in clean energy has escalated by 40%, driven not only by the importance of reducing emissions but also by other factors. The urgency for a transformation of the global energy system to align with the 1.5°C target is more compelling than ever, especially in light of August 2023 being the warmest August on record, following July 2023 as the hottest month ever recorded. The escalating frequency and severity of climate change impacts, coupled with increasingly dire scientific warnings, underscore the critical need for change [57].

In order to accomplish the most significant reductions in emissions by 2030, as outlined in the Net Zero Emissions by 2050 Scenario (NZE Scenario), it is crucial to increase the worldwide installed renewable capacity to 11,000 GW by 2030, which is three times the current capacity. Renewable power, namely generated from solar PV and wind sources, provides extensive accessibility, cost efficiency, and swift implementation. Even while it is anticipated that industrialized nations and China would be able to accomplish 85% of their contribution to the worldwide goal with the plans that they now have in place, other rising market and developing economies will need to implement improved policies and obtain assistance from the international community in order to achieve their own target [16].

1.4.5 The Paris Climate Agreement: A Catalyst for Change

The Paris Climate Agreement, which was adopted by 196 countries at the COP21 in Paris and came into force in November 2016, represents the global commitment to combat climate change. The document describes a collaborative endeavor to limit the rise in the Earth's average temperature to a level far lower than 2°C above the temperature before industrialization, with a particularly ambitious goal of 1.5°C. The recent focus of global leaders on the need of restricting warming to 1.5°C emphasizes the increasing urgency, as surpassing this barrier might lead to more frequent and severe climate-related disasters, as emphasized by the United Nations' Intergovernmental Panel on Climate Change [58,59].

To adhere to the 1.5°C target, it's imperative for greenhouse gas emissions to reach their peak before 2025 and undergo a 43% reduction by 2030. The Paris Agreement stands as a groundbreaking accord in the climate change dialogue, binding all nations to a collective fight against climate change and its ramifications, fostering a shift toward both economic and social transformation, grounded in scientific insights. The Agreement functions on a dynamic five-year cycle, promoting nations to gradually submit more aggressive national climate action plans, referred to as nationally determined contributions (NDCs) [60].

The COP27 requested the parties to review and improve their 2030 objectives under their NDCs by the end of 2023. This was done in response to the urgency to limit global warming to about 1.5°C. This request emphasizes the integration of equity principles and differentiated responsibilities, taking into account each nation's unique circumstances. This realignment aims to not only align with the Paris Agreement's temperature goals but also integrate climate action within broader objectives of sustainable development and poverty eradication [61].

NDCs serve as a platform for countries to outline their strategies for emission reduction and resilience building against climate impacts. In addition to these, long-term low greenhouse gas emission development strategies provide a forward-thinking outlook, situating NDCs within the context of long-term national planning and growth priorities. While long-term low greenhouse gas emission development strategies are not obligatory, they encapsulate the long-term aspirations that guide immediate actions outlined in the NDCs.

The Paris Agreement also lays a framework for support, encompassing financial, technical, and capacity building assistance for nations in need. Financial support is pivotal, especially for mitigation efforts requiring significant investments to curb emissions and for adaptation strategies to counteract climate change effects [62]. The advancement and sharing of technology are critical for boosting climate resilience and minimizing greenhouse gas emissions. The Agreement envisions a technology framework to steer the successful operation of the Technology Mechanism.

Capacity building is crucial for developing nations lacking the resources to tackle climate change challenges. The Agreement emphasizes the importance of increasing capacities in these countries, backed by developed nations' support. Furthermore, the enhanced transparency framework (ETF) set to commence in 2024 mandates comprehensive reporting on mitigation efforts, adaptation measures, and the support exchanged, fostering a transparent and accountable global climate action environment.

This collective progress will be assessed through the Global Stocktake, informing subsequent rounds of NDCs, thereby creating a cycle of continuous improvement and heightened ambition. Although the journey toward the Paris Agreement's objectives requires significant escalation in climate action, the advancements since its inception have already propelled low-carbon solutions and markets, showcasing the potential for a transformative shift toward a sustainable, carbon neutral future. By 2030, zero-carbon solutions are projected to be competitive in sectors responsible for over 70% of global emissions, presenting new opportunities for innovation and leadership in the clean energy shift [63].

1.4.6 Driving the Clean Energy Transition: Urgency and Action for Global Efficiency and Cooperation

The recent surge in global CO_2 emissions attained a high of 37 billion tons in 2022. While there's an increase, there is also a hopeful aspect: forecasts suggest that the requirement for coal, oil, and natural gas will attain its maximum in this decade because of the swift implementation of clean energy technologies, even without the introduction of new climate policies. This trend, while promising, falls short of the ambitious 1.5°C target set forth by international agreements [64].

In the past 2 years, the adoption of solar PV installations and EVs has aligned with the benchmarks outlined in the 2021 Net Zero by 2050 report. Governmental responses to the pandemic and the energy crisis catalyzed by geopolitical tensions have encouraged measures to accelerate the adoption of sustainable energy solutions, with the industry scaling up to meet the demand. If the announced expansions in manufacturing capacities for solar PV and batteries materialize, they could meet the demand projections for 2030, contributing significantly to the decarbonization efforts.

The possibility of limiting global warming to 1.5°C is still achievable, largely due to the proliferation of renewables, enhancements in energy efficiency, methane emission reductions, and broader electrification using current technologies. These measures account for over 80% of the necessary emission reductions by 2030. The expansion of clean energy, notably solar PV and wind, is instrumental in reducing fossil fuel demand by over 25% this decade, with well-crafted policies playing a crucial role in facilitating this transition [65].

The goal is to increase the worldwide installed capacity of renewable energy sources to 11,000 GW by 2030, which is three times the current capacity that represents the magnitude of effort needed to achieve emissions reduction targets. To achieve this objective, it is necessary to expedite the licensing procedures, enhance and update the power networks, tackle the difficulties in the supply chain, and safely include renewable energy sources that vary in their availability.

Enhancing the energy intensity improvement rate by 100% by 2030 would yield energy-saving equivalent equal to the present oil usage in road transport. This emphasizes the critical role of efficiency in lowering emissions, strengthening energy security, and improving affordability. At a global scale, these advancements will result from improving the technical efficiency of equipment, transitioning to more efficient energy sources, particularly electricity, and maximizing the utilization of energy and materials.

The electrification of various sectors, driven by technologies like EVs and heat pumps, is set to contribute significantly to emissions reductions [66]. By 2030, it is anticipated that electric car sales will constitute two-thirds of new car sales, and with the surge in heat pump sales exceeding expectations, the shift toward electrification is well underway [67].

Addressing methane emissions from the energy sector presents a cost-effective strategy to limit near-term global warming, with a 75% reduction by 2030 being both achievable and essential for meeting the 1.5°C target. The investment required for this reduction is minimal compared to the potential cost savings and the broader benefits of mitigating climate change [68].

While the momentum toward clean energy and reduced emissions is building, the need for concerted action, robust policies, and international cooperation has never been more critical. The transition to a net-zero emissions future necessitates a collective effort to implement the necessary measures, invest in clean energy infrastructure, and support the advancement and deployment of emerging technologies. This unified approach will not only pave the way for achieving the targets set by the Paris Agreement but also ensure a sustainable, secure, and equitable energy future for all.

1.5 CONCLUSION

This chapter highlights the urgent need for better energy efficiency and a quicker transition to sustainable energy to meet the growing global demand in a sustainable way. The analysis shows significant differences in energy use between developed and developing countries, underscoring the importance of customized energy policies and technologies. While reliance on fossil fuels remains a major hurdle, the rapid expansion of renewable energy provides a hopeful outlook. The case studies of various countries demonstrate different methods to improve energy efficiency and cut carbon emissions, reflecting their unique economic and environmental situations. International agreements like the Paris Climate Agreement play a key role in encouraging global cooperation and setting high targets for reducing emissions [69]. The chapter also highlights significant obstacles, such as the necessity for substantial investments in energy infrastructure, technological advancements, and effective policy enforcement. Despite these challenges, there are great opportunities for innovation and teamwork, paving the way for a sustainable and secure energy future. To accomplish this objective, it will be necessary for governments, industries, and communities worldwide to collaborate, emphasizing the interdependent nature of our global energy system.

REFERENCES

1. IEA (2021), Global Energy Review 2021, IEA, Paris, https://www.iea.org/reports/global-energy-review-2021, License: CC BY 4.0.
2. IEA (2024), Electricity 2024, IEA, Paris, https://www.iea.org/reports/electricity-2024, License: CC BY 4.0.
3. IEA (2023), Electricity Market Report 2023, IEA, Paris, https://www.iea.org/reports/electricity-market-report-2023, License: CC BY 4.0.

4. IEA (2021), Energy Efficiency 2021, IEA, Paris, https://www.iea.org/reports/energy-efficiency-2021, License: CC BY 4.0.
5. IEA, "Energy Efficiency Policy Database – Data & Statistics", https://www.iea.org/policies.
6. U.S. EIA. (2024). Total Energy Monthly Data, January 2024. DOE/EIA-0035(2024/1).
7. Tahmasebi, M. & Nassif, N. (2022). An intelligent approach to develop, assess and optimize energy consumption models for air-cooled chillers using machine learning algorithms. *American Journal of Engineering and Applied Sciences*, 15(3), 220–229, https://doi.org/10.3844/ajeassp.2022.220.229.
8. REN21. (2023). Renewables 2023 Global Status Report Collection, Economic & Social Value Creation. Paris: REN21 Secretariat. ISBN 978-3-948393-10-6.
9. Princeton University, "Princeton University Widens Net-Zero Goals and Lays out Dissociation Process to Advance Action on Climate Change", May 27, 2023, https://www.princeton.edu/news/2021/05/27/princeton-university-widens-net-zero-goalsand-lays-out-dissociation-process; Fossil Fuel Dissociation, https://fossilfueldissociation.princeton.edu.
10. 9 IEA, "Inflation Reduction Act of 2022 – Policies", https://www.iea.org/policies/16156-inflation-reduction-act-of-2022.
11. M. Barbanell, "A Brief Summary of the Climate and Energy Provisions of the Inflation Reduction Act of 2022", World Resources Institute, October 28, 2022, https://www.wri.org/update/brief-summary-climate-and-energy-provisions-inflationreduction-act-2022.
12. U.S. Energy Information Administration (EIA), "U.S. Energy Facts Explained," accessed May 29, 2024, https://www.eia.gov/energyexplained/us-energy-facts/.
13. U.S. Department of Energy, "Strategic Petroleum Reserve," accessed May 29, 2024, https://www.energy.gov/fe/services/petroleum-reserves/strategic-petroleum-reserve.
14. International Energy Agency (IEA), "World Energy Outlook 2022," IEA, Paris, https://www.iea.org/reports/world-energy-outlook-2022.
15. Baffes, J., Kose, M. A., Ohnsorge, F., & Stocker, M. (2015). The 2014–16 Oil Price Collapse in Retrospect: Sources and Implications. The World Bank. Retrieved from World Bank Documents (World Bank).
16. United Nations Framework Convention on Climate Change (UNFCCC), "The Paris Agreement," accessed May 29, 2024, https://unfccc.int/process-and-meetings/the-paris-agreement/the-paris-agreement.
17. International Renewable Energy Agency (IRENA), "Renewable Capacity Statistics 2024," IRENA, Abu Dhabi, https://www.irena.org/Statistics/View-Data-by-Topic/Capacity-and-Generation/Technologies.
18. Bloomberg NEF, "Clean Energy Investment Trends 2020," accessed May 29, 2024, https://about.bnef.com/clean-energy-investment-trends-2020.
19. Bloomberg NEF, "Energy Storage Investments to Reach $620 Billion by 2040," accessed May 29, 2024, https://about.bnef.com/blog/energy-storage-investments-to-reach-620-billion-by-2040/.
20. Our World in Data, "Carbon Intensity of Electricity", https://ourworldindata.org/grapher/carbon-intensity-electricity.
21. Chinese Ministry of Housing and Urban-Rural Development, "14th Five-Year' Building Energy Efficiency and Green Building Development Plan", 2021, https://www-mohurd-gov-cn.translate.goog/gongkai/fdzdgknr/zfhcxjsbwj/202203/20220311_765109.htm.
22. National Development and Reform Commission (NDRC), "14th Five-Year Plan," accessed May 29, 2024, https://www.ndrc.gov.cn/fzggw/jjjskfy/index.html.
23. International Energy Agency (IEA), "China Energy Outlook 2023," IEA, Paris, https://www.iea.org/reports/china-energy-outlook-2023.

24. Global CCS Institute, "CCUS in China: Market Trends and Opportunities," Global CCS Institute, accessed May 29, 2024, https://www.globalccsinstitute.com/resources/publications-reports-research/ccus-in-china-market-trends-and-opportunities/.
25. "China Power Sector Annual Investment Trends," accessed May 29, 2024, https://about.bnef.com/blog/china-power-sector-annual-investment-trends.
26. International Energy Agency (IEA), "India Energy Outlook 2023," IEA, Paris, https://www.iea.org/reports/india-energy-outlook-2023.
27. Government of India, "Panchamrit Strategy at COP26," Ministry of Environment, Forest and Climate Change, https://pib.gov.in/PressReleasePage.aspx?PRID=1776488.
28. Central Electricity Authority (CEA), "Monthly Installed Capacity Report," May 2023, https://cea.nic.in/monthly-installed-capacity-report/.
29. U.S. Department of Commerce, "U.S.-India Energy Cooperation," accessed May 29, 2024, https://www.commerce.gov/news/fact-sheets/2022/09/us-india-energy-cooperation.
30. Press Information Bureau, Government of India, "Government aims to install 250 million smart meters by 2025," accessed May 29, 2024, https://pib.gov.in/PressReleasePage.aspx?PRID=1757718.
31. Federal Ministry for Economic Affairs and Climate Action (BMWK), "Germany's Renewable Energy Investments," accessed May 29, 2024, https://www.bmwk.de/Redaktion/EN/Dossier/energy-transition.html.
32. Federal Environment Agency (UBA), "Germany's Greenhouse Gas Emissions 2022," accessed May 29, 2024, https://www.umweltbundesamt.de/en/topics/climate-energy/greenhouse-gas-emissions.
33. BMWK Federal Ministry for Economic Affairs and Climate Action, https://www.bmwk.de/Redaktion/EN/Dossier/energy-transition.html.
34. Federal Ministry for Economic Affairs and Climate Action (BMWK), "Germany's Energy Consumption 2022," accessed May 29, 2024, https://www.bmwk.de/Redaktion/EN/Dossier/energy-transition.html.
35. Federal Network Agency (Bundesnetzagentur), "Electricity Generation Data 2022," accessed May 29, 2024, https://www.bundesnetzagentur.de/EN/Home/home_node.html.
36. BMUV Federal Ministry for the Environment, Nature Conservation, Nuclear Safety and Consumer Protection, https://www.bmuv.de/en/.
37. UBA Federal Environment Agency, https://www.umweltbundesamt.de/en.
38. DENA German Energy Agency, https://www.dena.de/en/home/.
39. German National Hydrogen Strategy, "Hydrogen Demand Forecasts," accessed May 29, 2024, https://www.bmwi.de/Redaktion/EN/Artikel/Energy/national-hydrogen-strategy.html.
40. Federal Network Agency (Bundesnetzagentur), "LNG Imports and Hydrogen Transition," accessed May 29, 2024, https://www.bundesnetzagentur.de/EN/Home/home_node.html.
41. S. Treolar, "Norway Approves Over $18 billion of Oil, Natural Gas Projects", Bloomberg, June 28, 2023, https://worldoil.com/news/2023/6/28/norway-approves-over-18-billion-of-oil-natural-gasprojects.
42. Norwegian Water Resources and Energy Directorate (NVE), "Hydropower in Norway," accessed May 29, 2024, https://www.nve.no/energy/hydropower-in-norway/.
43. Norwegian Ministry of Petroleum and Energy, "Offshore Wind Power in Norway," accessed May 29, 2024, https://www.regjeringen.no/en/topics/energy/renewable-energy/offshore-wind/id2006130/.
44. Norwegian Electric Vehicle Association, "Electric Vehicle Market in Norway," accessed May 29, 2024, https://elbil.no/english/norwegian-ev-market/.
45. Norwegian Ministry of Petroleum and Energy, "Longship Carbon Capture and Storage Project," accessed May 29, 2024, https://www.regjeringen.no/en/topics/energy/longship-carbon-capture-and-storage-project/id2860073/.

46. The White House, "Executive Order on Catalyzing Clean Energy Industries and Jobs through Federal Sustainability," accessed May 29, 2024, https://www.whitehouse.gov/briefing-room/presidential-actions/2021/12/08/executive-order-on-catalyzing-clean-energy-industries-and-jobs-through-federal-sustainability/.
47. U.S. General Services Administration (GSA), "Federal Fleet Electrification," accessed May 29, 2024, https://www.gsa.gov/governmentwide-initiatives/sustainability/federal-fleet-electrification.
48. U.S. General Services Administration (GSA), "Net-Zero Emissions Procurement by 2050," accessed May 29, 2024, https://www.gsa.gov/governmentwide-initiatives/sustainability/net-zero-emissions-procurement.
49. U.S. Department of Energy (DOE), "Net-Zero Emissions Building Portfolio," accessed May 29, 2024, https://www.energy.gov/eere/buildings/net-zero-emissions-building-portfolio.
50. The White House, "Executive Order on Tackling the Climate Crisis at Home and Abroad," accessed May 29, 2024, https://www.whitehouse.gov/briefing-room/presidential-actions/2021/01/27/executive-order-on-tackling-the-climate-crisis-at-home-and-abroad/.
51. IRENA, "Renewable Energy Targets in 2022: A Guide to Design", November 2022, https://www.irena.org/Publications/2022/Nov/Renewable-energy-targets-in-2022.
52. International Renewable Energy Agency (IRENA), "Renewable Capacity Statistics 2021," accessed May 29, 2024, https://www.irena.org/publications/2021/March/Renewable-Capacity-Statistics-2021.
53. European Commission, "The European Green Deal," accessed May 29, 2024, https://ec.europa.eu/info/strategy/priorities-2019-2024/european-green-deal_en.
54. European Commission, "REPowerEU: Affordable, Secure and Sustainable Energy for Europe", May 18, 2022, https://commission.europa.eu/strategy-and-policy/priorities-2019–2024/european-green-deal/repowereu-affordable-secure-and sustainable-energy-europe_en.
55. World Meteorological Organization, "Climate and Weather Extremes in 2022 Show Need for More Action", December 23, 2022, https://public.wmo.int/en/media/news/climate-and-weather-extremes-2022-show-need-more-action.
56. World Meteorological Organization (WMO), "State of the Global Climate 2022," accessed May 29, 2024, https://public.wmo.int/en/our-mandate/climate/wmo-statement-state-of-global-climate. NASA, "July 2023 Hottest Month on Record," accessed May 29, 2024, https://climate.nasa.gov/news/3124/july-2023-hottest-month-on-record.
57. International Renewable Energy Agency (IRENA), "World Energy Transitions Outlook 2023," IRENA, Abu Dhabi, https://www.irena.org/publications/2023/Jun/World-Energy-Transitions-Outlook-2023.
58. Intergovernmental Panel on Climate Change (IPCC), "Global Warming of 1.5°C," accessed May 29, 2024, https://www.ipcc.ch/sr15/.
59. Intergovernmental Panel on Climate Change (IPCC), "Climate Change 2022: Mitigation of Climate Change," accessed May 29, 2024, https://www.ipcc.ch/report/ar6/wg3/.
60. United Nations, "Climate Change and Sustainable Development Goals," accessed May 29, 2024, https://www.un.org/sustainabledevelopment/climate-change/.
61. International Energy Agency (IEA), "Net Zero by 2050: A Roadmap for the Global Energy Sector," IEA, Paris, https://www.iea.org/reports/net-zero-by-2050.
62. 3 IMF Climate Finance Policy Unit, "ESG Monitor Q4 2021", February 8, 2022, https://www.imfconnect.org/content/dam/imf/News%20and%20Generic%20Content/GMM/Special%20Features/ESG%20Monitor%20Q4%202021.pdf.
63. International Energy Agency (IEA), "Global Energy Review: CO2 Emissions in 2022," IEA, Paris, https://www.iea.org/reports/global-energy-review-co2-emissions-in-2022.

64. International Energy Agency (IEA), "World Energy Outlook 2023," IEA, Paris, https://www.iea.org/reports/world-energy-outlook-2023.
65. International Energy Agency (IEA), "Global EV Outlook 2023," IEA, Paris, https://www.iea.org/reports/global-ev-outlook-2023.
66. McKinsey, "Renewable Development: Overcoming Talent Gaps", https://www.mckinsey.com/industries/electric-power-andnaturalgas/our-insights/renewable-energy-development-in-anet-zeroworld-overcoming-talent-gaps.
67. International Energy Agency (IEA), "Methane Tracker 2023," IEA, Paris, https://www.iea.org/reports/methane-tracker-2023.
68. Climate Action Tracker, "CAT Climate Target Update Tracker", https://climateaction-tracker.org/climate-target-update-tracker-2022
69. IRENA, "Accelerating Energy Transition Solutions at COP27", November 18, 2022, https://www.irena.org/News/articles/2022/Nov/Accelerating-energy-transition-solutions-at-COP27.

2 Innovations in Coal-Fired Power Plants

Shalini Kumari and Vicky Saurabh

2.1 INTRODUCTION

This chapter deals with different innovations in the energy sector, primarily from coal as the dominant energy source. Traditionally, coal is a gray energy source because it emits high carbon emissions into the environment. This chapter explains various strategies and innovations to transform coal energy sources from gray to green. The transformation from gray to green in coal energy sources will lead to a better understanding of the business's carbon and energy management practices.

2.1.1 Importance of Coal in Electricity Generation

In today's modern life, we cannot imagine our daily activities without electricity. From morning alarm to evening dinner, all the appliances we are using majorly depend on electricity. The dominant sources to generate electricity in the world are fossil fuels. According to the report of World Energy Resources (2022), fossil fuels account for more than 80% of the world's primary energy supply (IEA, 2022). Various fossil fuels, such as coal, natural gas, and petroleum, contribute to generating electricity. Among them, coal as the source is the dominant fossil fuel in generating electricity. Apart from fossil fuels, hydroelectricity and nuclear power also produce electricity, but the percentage seems to be less than 10%. Even for the next 10–20 years, the projected electricity production by hydro and nuclear seems to be the same. The limitations of hydroelectric power stem from the scarcity of suitable sites and their environmental impacts, while nuclear power faces constraints due to political and environmental concerns. The other options to generate electricity are renewable energy sources. Nowadays, renewable energy sources are gaining attention, but their contribution cannot fulfill consumers' electricity demand. Renewables such as solar and wind power are non-dispatchable and intermittent, leading to uncertainty in the energy market due to fluctuations in demand and supply. Moreover, supply-side issues such as dependency on geographical locations and climatic conditions further affect the reliability of renewable energy sources.

Figure 2.1 shows the percentage of contribution of supply of different energy sources (1971–2021) in achieving energy demand worldwide. The chart in Figure 2.1 shows the significant contribution in supplying electricity dependent on fossil fuels. However, renewable energy contributes 12%–15% of total energy supply. According to the report of World Energy Resource (2022), the projected data state that the different sources of fossil fuels such as crude oil and gas reserves are expected to diminish

DOI: 10.1201/9781003452072-2

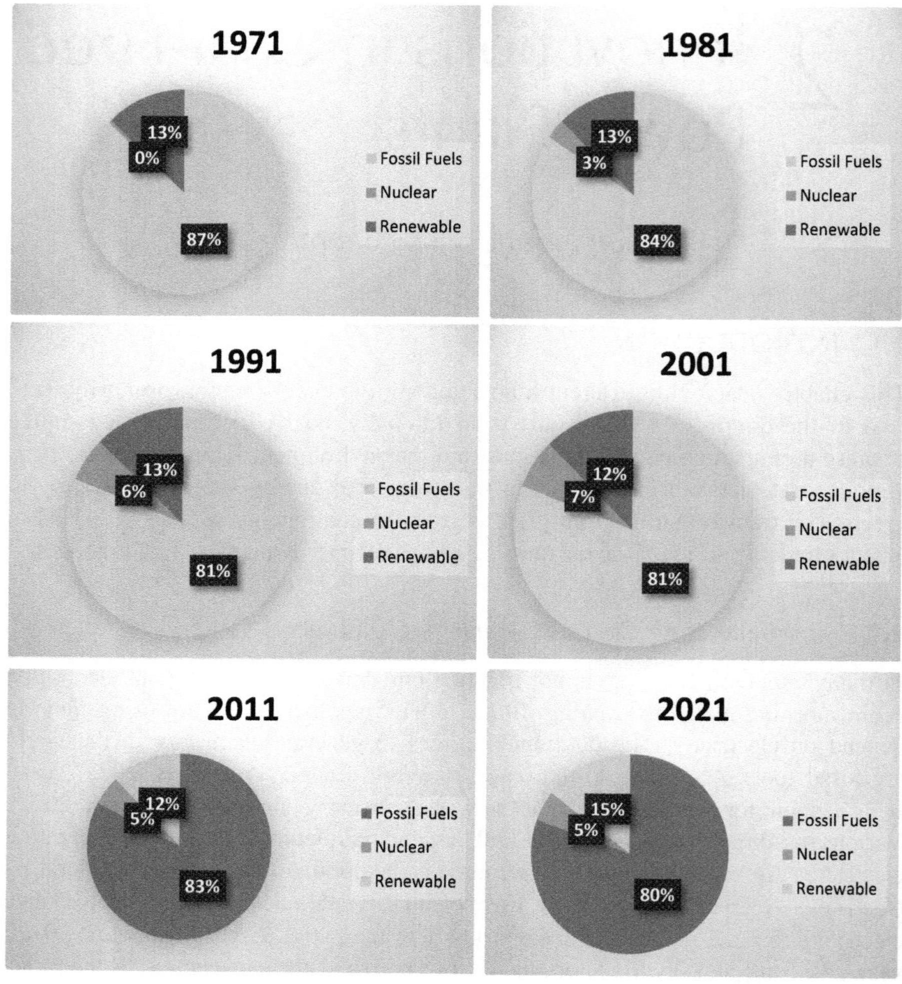

FIGURE 2.1 Total energy supply by resource 1971, 1981, 1991, 2001, 2021 (World Energy Resources, 2022).

within 4–6 decades, and the only available possible fossil fuel sources would be coal for over a century. This extended longevity, widespread distribution, and affordability, particularly benefiting developing nations, solidify coal's prominent position in the global energy mix.

Despite the rapid deployment of renewable energy technologies and ongoing discussions surrounding climate change, coal has experienced the most substantial increase in energy demand compared to other sources. This surge can be attributed to coal's reliability, abundant availability, low cost, and immunity to price volatility, unlike oil and natural gas. Events such as the "oil price spike" in 2008, where oil prices increased to $150/barrel, and the subsequent decline in oil and gas prices in 2015 show the inherent volatility in the fossil fuel market, especially in the natural gases and oil. This volatility is likely to persist due to geopolitical instability in significant oil and gas-producing regions and the continued production of shale gas.

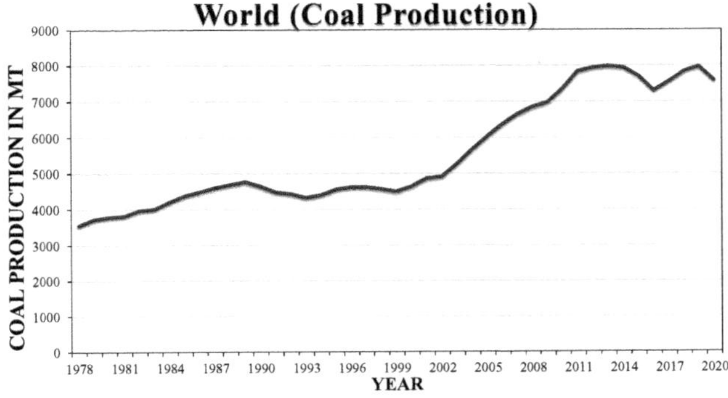

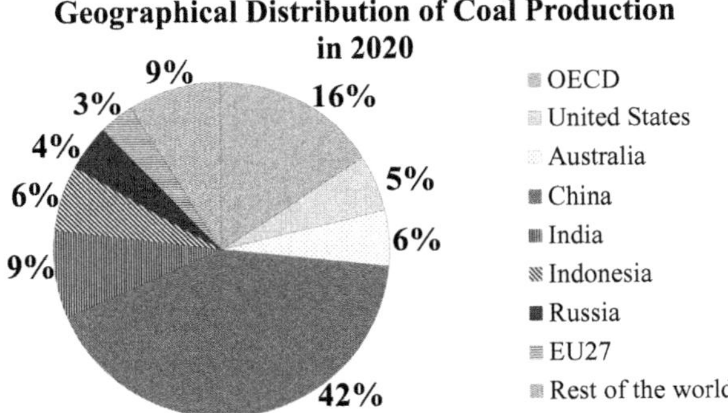

FIGURE 2.2 Total production of coal in the world (IEA, World Total Coal Production).

2.1.2 Global Coal Production Outlook

The global coal production consistently exhibits an upward trend, as illustrated in Figure 2.2. This increasing trend reflects the world's reliance on coal for electricity generation due to its affordability as an energy source. Geographically, approximately 42% of the world's total coal production originates from China as of 2020, indicating a significant contribution. Other major coal-producing nations include India and the United States of America.

2.1.3 The Current Landscape of Coal Production in India

India heavily relies on coal as its primary energy source and ranks as the third largest coal producer globally. The country has abundant domestic coal reserves and convenient access to affordable imported coal through its numerous coastal ports. Due to the availability of coal reserves, India's rapid economic growth depends on coal. This growth further increases demand for electricity and materials for infrastructure development, such as steel and cement. Coal plays a pivotal role in electricity generation and the production of these essential materials, highlighting its importance in India's economic expansion and vice versa.

Despite India's economic progress, approximately a quarter of its population still grapples with energy poverty. The power sector in India faces a significant demand–supply

TABLE 2.1
Details of Coal Reserves in India as of 2022 (Ministry of Coal)

State of India	Total Resource (Million Tons)
Odisha	88,104.60
Jharkhand	86,660.10
Chhattisgarh	74,191.76
West Bengal	33,871.25
Madhya Pradesh	30,916.73
Telangana	23,034.20
Maharashtra	13,220.71
Bihar	4,437.18
Andhra Pradesh	4,141.87
Uttar Pradesh	1,061.80
Meghalaya	576.48
Assam	525.01
Nagaland	478.31
Sikkim	101.23
Arunachal Pradesh	90.23
Total	**361,411.46**

gap of over 10%. India has planned substantial expansion in power generation capacity to address this issue. While various energy sources will contribute to this growth, coal is expected to meet a significant portion of the future energy demand in India.

Table 2.1 outlines India's total coal resources, with significant deposits concentrated in eastern states such as Andhra Pradesh, Arunachal Pradesh, Chhattisgarh, Jharkhand, Odisha, and West Bengal. However, Indian coal faces challenges such as low rank (lignite and sub-bituminous), high ash content (up to 40% in some coal types), and depth of extraction exceeding 300 m in many cases. Consequently, India relies on substantial imports of high-quality coal from countries such as Australia, China, Indonesia, and South Africa, while exporting a small fraction to neighboring nations such as Bangladesh, Bhutan, and Nepal.

2.2 COAL-FIRED POWER PLANTS

Power plants are complex industrial facilities that use generators to convert fuel into electricity. They are primarily categorized into two groups based on their source: conventional power plants and non-conventional power plants. Conventional power plants include thermal, diesel, gas, hydro, and nuclear. Non-traditional power plants encompass solar, wind, tidal, and geothermal energy sources. Thermal power plants, particularly coal-fired plants, employ pulverized coal (PC) combustion to generate high-pressure steam, which drives steam turbines (STGs) to produce electricity from mechanical energy. These plants have evolved with technological advancements, developing subcritical, supercritical, ultra-supercritical, or advanced ultra-supercritical PC power plants. However, coal-fired plants are known for their high emissions and relatively low efficiency, which are significant environmental concerns.

TABLE 2.2
Comparative Result of PC and IGCC Power Plants

	PC Power Plant		
	Subcritical Boiler	Supercritical Boiler	IGCC
Gross process output (kW)	583,315	580,260	770,350
Auxiliary power requirement (kW)	32,870 (5.6%)	30,110 (5.18%)	130,100 (16.89%)
Net power output (kW)	550,445	550,150	640,250
Coal flow rate (lb/hr)	437,699	646,589	489,634
HHV thermal input (kW)	1,496,479	1,406,161	1,674,044
Net plant HHV efficiency (%)	36.8%	39.1%	38.2%
Net plant Higher heating value (HHV) heat rate (Btu/kW-hr)	9,276	8,721	8,922
Raw water usage, gpm	6,212	5,441	4,003
Total plant cost ($*1,000)	852,612	866,391	1,160,919
Total plant cost ($/kW)	1,549	1,575	1,813
LCOE (mills/kWh)	64	63.3	78
CO_2 emissions (lb/hr)	1,038,110	975,370	1,123,781

On the other hand, integrated gasification combined cycle (IGCC) power plants utilize coal for electricity production through gasification. In an IGCC plant, coal is gasified, followed by cleaning the produced syngas before being used in combustion turbine generators (CTGs), heat recovery steam generators (HRSGs), and STGs to generate electricity. These plants incorporate an air separation unit (ASU) to provide the oxygen necessary for coal gasification, impacting the overall cost of the power plant. IGCC plants offer advantages such as lower CO_2 emissions and relatively higher efficiency compared to traditional PC plants, making them an attractive option for cleaner electricity generation. These advancements reflect ongoing efforts in the power generation sector to improve efficiency and reduce environmental impact. Table 2.2 shows the comparative result between PC and IGCC power plants and its difference in terms of net plant efficiency, levelized cost of electricity (LCOE), and carbon emissions (NETL Report, 2008).

2.2.1 Energy Consumption Structure in a Coal-Fired Plants

In coal-fired power plants, energy is used to carry out various processes involved in generating electricity. The energy lost in these operations is referred to as the auxiliary energy requirement. This energy is mainly used in operations such as:

- **Coal Handling and Conveying System (0.07% Energy Requirement):** The coal handling and conveying system in coal-fired power plants is responsible for receiving, processing, storing, and feeding coal bunkers. This process consumes a notable amount of energy.
- **Limestone Handling and Reagent Preparation (0.16% Energy Requirement):** The primary purpose of limestone is to react with sulfur dioxide, thereby reducing emissions of sulfur, a component found in coal.

- **Pulverizers (0.5% Energy Requirement):** The bulk of the energy used by pulverizers goes toward three main tasks: pulverization, drying, and classification. Their primary function is to deliver fine coal particles to the boiler.
- **Ash Handling (0.098% Energy Requirement):** This mechanical system collects, manages, and disposes of ash from burning coal. There are generally two types of ash: bottom ash, which makes up about 20% of the total ash, and Fly Ash, which accounts for the remaining 80%. Fly ash is also collected in the baghouse.
- **Boiler Operation (1.84% Energy Requirement):** Coal is burned in the boiler to produce electricity. The boiler uses primary air fans, forced air fans, and induced air fans to ensure adequate air supply and proper combustion.
- **Selective Catalytic Reduction (SCR) (0.01% Energy Requirement):** Energy is used to separate nitrogen oxide emissions from the boiler.
- **Pumps (1.3% Energy Requirement):** Pumps in power plants serve various purposes, including flue gas desulfurization (FGD) pumps and agitators, condensate pumps, steam turbine auxiliaries (which convert steam energy into mechanical energy for electricity generation), and circulating water pumps.
- **Cooling Tower Fans (0.48% Energy Requirement):** Energy is consumed in operating cooling towers, dissipating excess heat from the power plant's systems.
- **Transformer Loss and Miscellaneous (0.65% Energy Requirement):** Energy is also used for transformer losses and other miscellaneous needs.

These areas represent significant stages in the energy consumption structure of a coal-fired power plant and are pictorially represented in Figure 2.3.

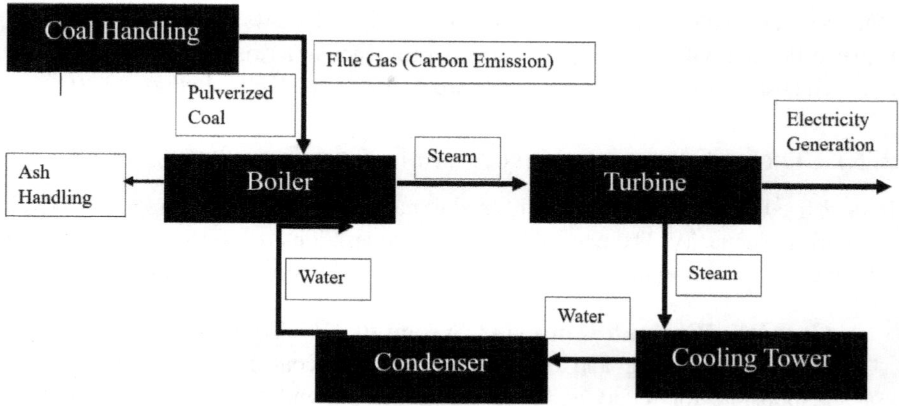

FIGURE 2.3 Basic framework of electricity generation from coal-fired power plant.

2.3 INNOVATIONS IN COAL-FIRED POWER PLANTS: THE CONCEPT OF CARBON ENERGY MANAGEMENT

Carbon and energy management techniques in industries have become essential in addressing the challenges of climate change and global warming, mainly stemming from emission-intensive sectors. The concept of carbon management gained prominence with the widespread use of the term "carbon footprint" (Wiedmann and Minx, 2008), highlighting the necessity to minimize carbon emissions.

The first oil crisis in 1973 served as the impetus for adopting energy management techniques during the second oil crisis in 1979, which saw a sharp increase in energy costs. (Krapels, 1980). Energy management initially focused on improving commercial and industrial energy efficiency.

The evolution of energy management is documented in various historical periods of literature. Phase 1 (from 1973 to 1981) emphasized "save it" strategies, focusing on energy conservation (O'Callaghan and Probert, 1977). The objective of Phase 2 of energy management (from 1981 to 1993) is "managing it," including initiatives like privatizing the electricity supply sector (Kumar et al., 1984; Webb, 1985; O'Riordan, 1989; Ranganathan, 1993).

The subsequent phase, "Purchase it," lasted from 1993 to 2000, when environmental concerns gained prominence and energy management became integrated into business strategies by introducing environmental programs (Wiser, 1998; Harris and Johnson, 2000). This era marked the emergence of carbon and energy management practices.

Phase 4, known as the "Respond to Climate Change Levy," spanned from 2000 to 2010, focusing on addressing climate change impacts and implementing measures to reduce carbon emissions (Kannan and Boie, 2003; Capehart et al., 2003).

Phases 3 and 4 address the issues of carbon emissions. Thus, introducing the Kyoto Protocol in 1997 (Protocol, 1997) is a crucial attempt to tackle the growing environmental challenges. This protocol made it mandatory for member countries to work toward reducing emissions, with a binding agreement placing the primary responsibility on developed nations due to their higher energy demands and consequent carbon emissions into the atmosphere.

Implemented in 2005 (Protocol, 2005), the Kyoto Protocol emphasized the "common but differentiated responsibility principle," highlighting the heavier burden on developed nations. Countries that committed to emission reductions were required to achieve specified percentage reductions within 5 years from 2008, based on the 1990 emission levels. Due to escalating energy consumption and carbon dioxide emissions, a stricter policy was introduced under the Kyoto Protocol, mandating signatory nations to cut emissions by a minimum of 18% from 1990 levels within 8 years, from 2013 to 2020.

From around 2010 onward, there was a notable shift toward advanced clean technologies and a greater reliance on renewable energy sources. This shift marked an evolution in energy management toward carbon energy management, with a renewed focus on mitigating the adverse effects of climate change. Numerous studies and initiatives have contributed to this shift, emphasizing the importance of adopting sustainable practices and reducing carbon footprints (Abdelaziz et al., 2011; Bunse et al., 2011; Ates and Durakbasa, 2012; Schulze et al., 2016).

Carbon and energy management can overcome the variegated responses and barriers (economic, organizational, and behavioral) to society regarding climate change (Ates and Durakbasa, 2012; Backlund et al., 2012). There are four practical goals for effective carbon and energy management. Firstly, comprehend the work of climate systems and understand the predictable impact and the key drivers that induce anthropogenic climate change. Secondly, calculate and forecast the emission release by country and source and assimilate the knowledge of carbon inventory. Thirdly, understand the offsetting principles and process and find the best feasible decision analysis model out of various control strategies. Lastly, understand the existing climate and energy policies and develop the policies based on the best feasible decision model.

The third objective of carbon and energy management is one of the mitigating options for reducing anthropogenic carbon emissions from the traditional coal-fired power plant. Carbon capture and storage (CCS) is one of the mitigating options that fulfill the third objective of carbon and energy management.

2.3.1 Carbon Capture and Storage (CCS)

CCS refers to the capture of CO_2 from a specific source and its subsequent transportation to a selected storage place to effectively manage and reduce global carbon emissions. The point source of pollution might originate from coal-fired power plants, cement and steel industries, as well as other sources associated with energy production. According to the IPCC (Intergovernmental Panel on Climate Change), by 2050, CCS will be a highly secure technology capable of reducing almost 20% of global carbon emissions. CCS typically comprises three main steps: capture, transport, and storage (Mathieu, 2006). Carbon capture is viable from many stationary sources rather than mobile or tiny stationary sources. Carbon capture can be achieved using post-combustion, pre-combustion, and oxy-fuel techniques, as explained in the subsequent paragraph.

The paragraph below outlines three techniques for capturing carbon from major point sources.

2.3.2 Types of Carbon Capture (Wilberforce et al., 2019)

1. **Post-combustion Carbon Capture**: When the coal and air undergo the combustion process, the resultant is power, heat, carbon dioxide, and other gases. The heat generated produces steam in the boiler and drives the turbine to generate electricity. The carbon dioxide (present at dilute concentration and low pressure) captured from the flue gases after the combustion is the post-combustion carbon capture. Currently, there are three ways to capture the carbon from post-combustion methods: membrane-based CO_2 capture (permeable material and semi-permeable material), solvent CO_2 capture (chemical adsorption and physical adsorption), and sorbent CO_2 capture (chemical absorption and physical absorption). Figure 2.4 explains the post-combustion carbon capture method.

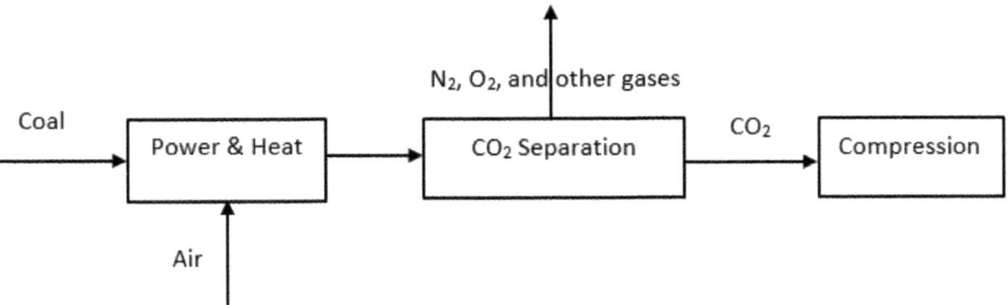

FIGURE 2.4 Capture of carbon dioxide by post-combustion method.

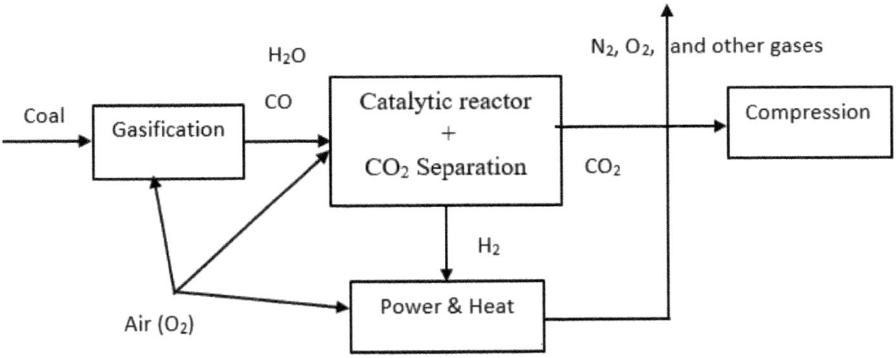

FIGURE 2.5 Capture of carbon dioxide by pre-combustion method.

2. **Pre-combustion Carbon Capture:** When the coal reacts with limited oxygen, it produces synthesis gas or syngas. The syngas comprises hydrogen, carbon monoxide, and other gaseous constituents. The syngas then passes through the catalytic reactor (water–gas shift reactor), where carbon monoxide converts into carbon dioxide (high concentration and pressure) and hydrogen. Pure hydrogen, as fuel, drives the turbine and generates the electricity. Generally, the physical and chemical absorption processes capture the carbon dioxide in pre-combustion carbon capture. Figure 2.5 explains the pre-combustion carbon capture method.
3. **Oxy-Fuel Combustion Capture method:** Figure 2.6 explains the oxy-fuel combustion capture method, where coal is burned with pure oxygen to generate heat and power. An air separator obtains pure oxygen. The coal burned produces high concentrations of CO_2 in the flue gases. The flue gases then pass through a CO_2 separator, such as the chemical looping cycle, where carbon gets captured before being released into the atmosphere.

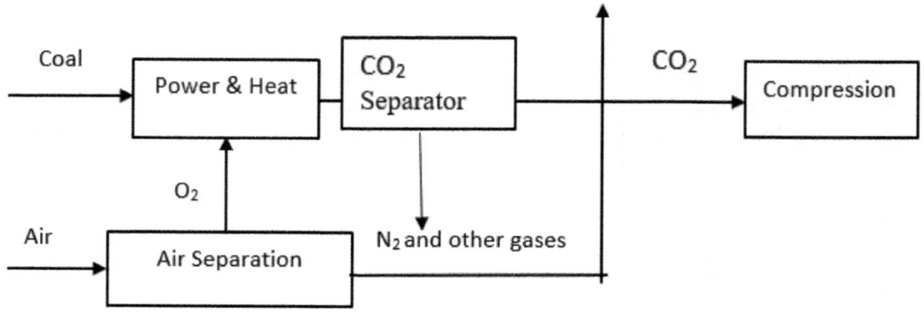

FIGURE 2.6 Capture of carbon dioxide by oxy-fuel combustion method.

2.3.3 Transportation and Storage

Transportation is crucial in the CCS process, linking the emission source to the storage site. The emission source typically refers to significant stationary emission points, while the storage site is the sink. This step is not entirely novel, as natural gas and petroleum transportation are already common, facilitated through pipelines or marine vessels. Carbon dioxide can be transported in several forms: gas, liquid, and solid. Gaseous CO_2 can be moved at near-atmospheric pressures, but it occupies a large volume, necessitating compression for efficient pipeline transport. Liquid CO_2, which takes up less space, can be transported by ships. However, solidifying CO_2 is typically not practical due to the significant energy required, leading to a substantial increase in costs. After capture and transportation, the final step in CCS is carbon dioxide storage, aimed at preventing carbon from entering the atmosphere and helping to regulate Earth's temperature. CO_2 storage can be achieved geologically or in the ocean.

2.3.4 Cost–Benefit Analysis of CCS

The cost–benefit analysis (CBA) evaluates different types of technology based on their costs and benefits. The benefits can be environmentally and financially related. The CBA is helpful for the decision-maker to decide on the adoption of the technology. It also facilitates the effective distribution of resources to the citizens. CBA is a labor-intensive and expertise-driven methodology that contributes to the overall expenses of the company. The objective of this section is to assess the CCS technology using CBA methodologies in order to determine the economic, social, and environmental impacts of a project. The costs associated with carbon capture include capital expenses, extra product costs, costs of carbon avoidance, and costs of carbon capture or removal.

2.3.5 Carbon Capture Cost

1. **Capital Cost:** The capital cost relates to the cost associated with installing technology, which is fixed in nature. The capital cost associated with carbon capture technology includes designing, procuring, and installing a carbon capture system. It can be denoted as the cost per kWh in normalized form. Also, the additional costs incurred due to the presence or integration of a carbon capture system in power plants are accounted as capital costs. The increase in capital cost of carbon capture can be determined by

comparing electricity production costs with and without a carbon capture unit. Considering uncertainty is crucial in calculating capital costs to account for the risks and uncertainties inherent in the dynamic business environment.

2. **Incremental Product Cost:** Product cost is the cost of producing electricity and carbon emissions from the power plant. Electricity is the primary product of power plants. However, the increase in product cost may be calculated by comparing the cost of energy generation with and without carbon capture equipment. Since power plants significantly contribute to carbon emissions, assessing how carbon capture systems affect electricity costs is essential. The factors affecting the product cost include the plant's operating life, per unit cost of fuel required, variable operating cost, and cost of electricity. The product cost is variable and fluctuates every year.
3. **Cost of Carbon Avoided:** The cost of carbon dioxide avoided for carbon capture technology is determined by the ratio of LCOE by carbon emission incurred with or without carbon capture technology. The LCOE incorporates the cost of capturing, transporting, and storing carbon dioxide in the context of CCS technology. This metric impacts the incremental cost of carbon capture for the particular type of plant.

$$\text{Cost of CO}_2\text{ avoided} = \frac{\text{LCOE}_{\text{with capture}} - \text{LCOE}_{\text{without capture}}}{\text{Emission of CO}_{2\,\text{without capture}} - \text{Emission of CO}_{2\,\text{with capture}}}$$

4. **Cost of Carbon Captured or Removed:** The carbon capture or removal cost is an additional expense linked to carbon capture technology. The cost of carbon capture is calculated by determining the ratio between the difference in the LCOE and the quantity of carbon dioxide that is removed. The cost of carbon capture serves as a tool for evaluating the economic feasibility of carbon capture technology when carbon is exchanged in the market. Assume that the market price for carbon collected is equal to the additional cost of generating power using the carbon capture equipment. Under those circumstances, a carbon capture system proves to be more advantageous as it produces energy at a comparable cost to the reference plant while exhibiting comparatively low carbon emissions.

$$\text{Cost of CO}_2\text{ captured/removed} = \frac{\text{LCOE}_{\text{with capture}} - \text{LCOE}_{\text{without capture}}}{\text{CO}_{2\,\text{removed}}}$$

2.3.6 Transportation Cost of Captured Carbon

After capturing carbon, it must be transported, which incurs transportation costs. The transportation of captured carbon can occur via pipeline or marine transport. Pipeline transportation costs encompass construction, operation, maintenance, and other expenses. Construction costs depend on factors such as the amount of carbon to be transported, distance to the storage area, and pipeline material. Labor is a significant component of construction costs. Operation and maintenance costs include monitoring expenses to ensure pipeline integrity and regular maintenance to prevent deterioration. Other costs include insurance, project management, design, right-of-way, contingencies, and regulatory fees.

Onshore pipelines on land may incur increased costs due to terrain, urbanization, and safety measures. Offshore pipelines, laid on the seabed, are more expensive due to higher pressure and lower temperature requirements. The size of the project influences costs, with smaller projects typically incurring higher costs. The cost of pipeline and marine transportation includes recurring expenses such as labor, fuel, electricity, port fees, and maintenance.

After transportation, carbon must be stored in geological or oceanic sinks. Ocean storage costs involve transport, handling, and offshore activities, while geological storage costs include capital, operating, and monitoring expenses.

Moving on to benefits, they can be categorized as non-financial and financial. Non-financial benefits include greenhouse gas reduction, contributions to sustainable development, and improvements in citizen welfare. Financial benefits include sustainable business growth, government incentives such as tax benefits and funding, carbon trading opportunities, and reduced electricity costs over time as CCS technology advances and becomes more widely adopted.

2.4 INNOVATIONS IN CLEANER ENERGY FROM COAL

Innovations in cleaner energy from coal have become increasingly crucial in addressing climate change due to global warming. The world has recognized the urgency of mitigating climate change. It has set targets to address this issue, with low-emission coal technologies playing a vital role in meeting these targets. As the demand for coal continues to rise, the importance of clean coal technologies becomes even more pronounced. Traditionally, efforts to reduce carbon emissions from coal-based power plants have primarily focused on two approaches: the utilization of highly efficient technologies and the implementation of carbon capture, use, and storage (CCUS) methods. However, besides these established methods, there has been a growing interest in exploring unconventional and innovative energy solutions. This exploration has led to the development of diverse, sophisticated, clean coal technologies to enhance efficiency and address ecological concerns associated with coal utilization.

Some of the notable innovations in clean coal technology include:

Supercritical and Ultra-Supercritical PC Combustion: This method of producing electricity uses supercritical steam conditions to achieve high conversion efficiencies while operating at high pressure and temperature.

Integrated Gasification Combined Cycle (IGCC): This technique combines the combustion of coal gas in a gas turbine with the integration of coal gasification. It also concentrates on recovering waste heat in a boiler using a steam turbine.

Direct Coal-Fired Combined Cycle: This method recovers waste heat by combining a gas and steam turbine. It is comparable to IGCC. Instead of gasifying coal, it burns coal straight in the gas turbine.

Pressurized Pulverized Coal Combustion (PCC): This method uses a combined cycle based on coal, burning the coal at temperatures up to 20 bar and at a temperature of around 1,500°C. The technique is noteworthy for its ability to remove ash at high process temperatures and pressures, which permits the direct use of flue gas as a gas turbine input.

Underground Coal Gasification (UCG): UCG is the process of gasifying coal under in situ circumstances so that existing technology can be used to utilize inaccessible coal reserves.

2.4.1 Underground Coal Gasification

UCG is a process that involves the gasification of coal in situ to produce synthesis gas (syngas), as presented in Figure 2.7. The operating life of a UCG operation typically involves four main steps:

1. **Well Construction and Linkage:** To enable the injection of oxidants and the extraction of product gas, wells are drilled into the coal seam. These wells are strategically interconnected or extended to create an inseam channel. This network of channels is crucial for several reasons: It allows for the efficient injection of oxidants, which are necessary for the chemical reactions that generate syngas. Additionally, the interconnected channels support the development of cavities within the coal seam, which in turn facilitates the smooth flow and extraction of syngas. This integrated system ensures the effective operation and productivity of the gasification process.
2. **Ignition:** Once the wells are prepared, the coal seam is dried and ignited to initiate gasification.

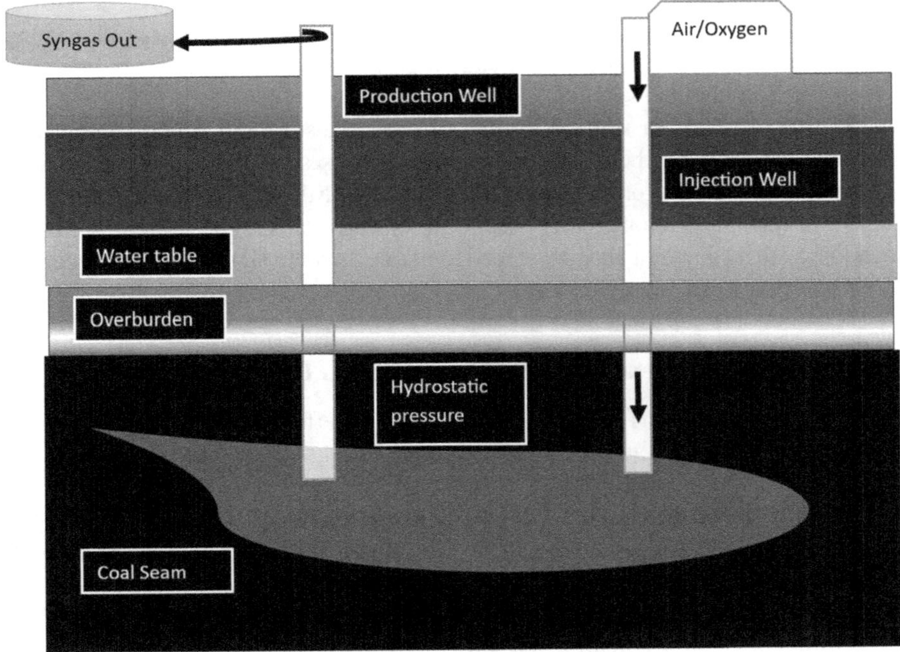

FIGURE 2.7 Framework of underground coal gasification.

3. **Gas Production:** At this stage, syngas is produced by a sequence of combustion and gasification events. At first, combustion processes occur, generating heat, carbon dioxide, and a tiny quantity of syngas by partially burning the coal. Subsequently, gasification processes take place, which involve the interplay of heat, carbon dioxide generated during the combustion process, pressure, steam, and carbon derived from the coal. These reactions result in a greater volume of syngas being produced. The syngas produced is subsequently transported from the gasification zone to the gas production well via pre-existing or man-made horizontal conduits. Upon reaching the surface, the syngas is subjected to further processing in order to purify and make it suitable for practical use.
4. **Decommissioning:** The gasification process is stopped according to established shutdown procedures once the available coal has been fully converted into gaseous products. The process typically involves shutting down injection and extraction wells and securing the site by safety and environmental regulations.

Overall, the UCG process offers a method for accessing coal reserves that are not economically viable using traditional mining techniques. However, proper management and monitoring throughout the operational phases are essential to ensure safety, environmental protection, and efficient gas production.

UCG represents a significant area of interest in efficiently utilizing coal resources that are typically inaccessible using current mining technologies. This process enables access to coal reserves beyond what is physically reachable through conventional mining methods, thus expanding the potential global coal resource. According to the World Energy Council, UCG can augment world coal reserves by approximately 600 billion tons.

UCG involves the in situ gasification of coal, transforming it into combustible gases. These gases can be utilized for various purposes, such as power generation, known as UCG–IGCC, or as a source for producing hydrogen, synthetic natural gas (SNG), or methanol, among other uses. Consequently, UCG represents a promising technology that can be further developed to harness a more significant portion of the world's largest fossil energy resource.

2.4.2 Unlocking Benefits: Advantages of UCG Process

UCG presents numerous advantages, such as cost-effectiveness, safety improvements, environmental benefits, and access to previously unminable coal deposits.

- **Low Operating Costs:** UCG boasts lower operating costs than conventional mining methods due to no transportation costs, minimal reactor maintenance, and reduced steam requirements.
- **Safety:** UCG eliminates many safety concerns associated with traditional mining operations, making it a safer option for coal extraction.
- **Exploitation of Unminable Deposits**: UCG enables the exploitation of coal reserves located deep underground that are otherwise inaccessible using current mining technologies.

- **Environmental Benefits:** Gasification in UCG is inherently cleaner, reducing particulate pollution compared to other coal utilization methods.
- **Carbon Dioxide Sequestration:** UCG offers the potential to store carbon dioxide generated during the process and utilize syngas in UCG cavities, contributing to carbon sequestration efforts.

2.4.3 Environmental and Commercial Challenges in UCG

- **Groundwater Contamination:** Gases produced during UCG can contaminate nearby aquifers, necessitating careful site selection and regular water quality monitoring to prevent damage.
- **Surface Subsidence:** Widening cavities in UCG operations increase the risk of roof collapse and subsequent subsidence, influenced by mechanical properties and thermal stresses due to high temperatures.
- **CO_2 Emissions:** Significant CO_2 production in UCG requires considerations for separation and utilization or storage to minimize atmospheric emissions, with the option of storing CO_2 in the UCG cavity.

2.4.3.1 Challenges to Commercialization

- **Limited Suitable Sites:** Adverse geological and hydrological conditions limit suitable coal seams for UCG, with improper site selection risking aquifer contamination and subsidence.
- **Low Controllability:** The inherent complexity of UCG makes controlling water influx, gas flow, and spalling rates challenging compared to surface gasifiers.
- **Uncertainties in Economics**: The model of UCG is still in development. The lack of robust, comprehensive modeling of UCG and unreliable field-scale data results in uncertainties in the economies.
- **Technical Uncertainties:** Obtaining reliable data is difficult and expensive, hindering a thorough understanding of UCG's fundamental processes and overall performance.
- **Public Perception:** Limited public awareness and perception of UCG present non-technical barriers to commercialization.

Addressing these challenges requires further research, inclusive decision-making processes, remote site selection, and developing confidence in UCG's efficiency and environmental safety, especially for deeper coal seams.

2.5 SUMMARY

Coal is an important source of electricity generation in the present situation. The dependency of fulfilling the electricity demand is mainly on coal. The total coal reserves in India are around 361,411.46 million tons, according to the 2022 report. This chapter details various technological advancements in coal-fired power plants, transforming the gray energy source into green energy sources. It discusses different types of power plants, mainly concentrating on PC and IGCC power plants. The PC power plant is further divided into subcritical and supercritical power plants. The comparative study between PC and IGCC is shown in this chapter.

Further, this chapter also discusses how carbon capture technology can retrofit the traditional coal-fired power plant; the cost analysis of carbon capture; UCG, one of the substitutes for PC power plants; and the challenges associated with generating electricity through UCG, including environmental impacts such as greenhouse gas emissions, air pollution, and water contamination. It also touches upon the economic and social considerations related to coal mining, energy security, and global climate change implications.

Overall, this chapter provides a comprehensive overview of coal's importance in electricity generation, highlighting its contributions and the continual attempts to improve its sustainability and mitigate its environmental impacts within framework of evolving energy landscapes and climate change concerns.

REFERENCES

Abdelaziz, E. A., Saidur, R., & Mekhilef, S. (2011). A review on energy saving strategies in industrial sector. *Renewable and Sustainable Energy Reviews*, 15(1), 150–168.

Ates, S. A., & Durakbasa, N. M. (2012). Evaluation of corporate energy management practices of energy-intensive industries in Turkey. *Energy*, 45, 81–91.

Backlund, S., Broberg, S., Ottosson, M., & Thollander, P. (2012). Energy efficiency potentials and energy management practices in Swedish firms. In: *Proceedings of the ECEEE Summer Study on Energy Efficiency in Industry,* Arnhem, Netherlands (pp. 669–677).

Bunse, K., Vodicka, M., Schönsleben, P., Brülhart, M., & Ernst, F. O. (2011). Integrating energy efficiency performance in production management–gap analysis between industrial needs and scientific literature. *Journal of Cleaner Production*, 19(6–7), 667–679.

Capehart, B. L., Turner, W. C., & Kennedy, W. J. (2003). *Guide to Energy Management.* The Fairmont Press, Inc, Lilburn, GA.

IEA (2022), World Energy Outlook 2022, IEA, Paris. https://www.iea.org/reports/world-energy-outlook-2022, Licence: CC BY 4.0 (report); CC BY NC SA 4.0 (Annex A).

Kannan, R., & Boie, W. (2003). Energy management practices in SME—case study of a bakery in Germany. *Energy Conversion and Management*, 44(6), 945–959.

Krapels, E. N. (1980). *Oil Crisis Management: Strategic Stockpiling for International Security.* Johns Hopkins University Press, Baltimore, London.

Kumar, A. R., Hackett, D. F., Eisenhauer, J., Vemuri, S., & Lugtu, R. (1984). Fuel resource scheduling, Part I-Overview of an energy management problem. *IEEE Transactions on Power Apparatus and Systems*, 7, 1542–1548.

Mathieu, P. (2006). The IPCC special report on carbon dioxide capture and storage. In *ECOS 2006- Proceedings of the 19th International Conference on Efficiency, Cost, Optimization, Simulation and Environmental Impact of Energy Systems* (pp. 1611–1618). National Technical University of Athens.

NETL (2008). https://www.netl.doe.gov/sites/default/files/netl-file/ASME-FY08-Fuel-Cell-Peer-Review-Report_Final.pdf.

O'Callaghan, P. W., & Probert, S. D. (1977). Energy management. *Applied Energy*, 3(2), 127–138.

O'Riordan, T. (1989). Electricity privatization and environmental accountability. *Energy Policy*, 17(2), 141–148.

Protocol, K. (1997). United Nations framework convention on climate change. *Kyoto Protocol, Kyoto*, 19, 497.

Protocol, K. (2005). Status of ratification. In *United Nations Framework Convention on Climate Change*. https://unfccc.int/process/the-kyoto-protocol/status-of-ratification#:~:text=The%20Kyoto%20Protocol%20entered%20into,of%20the%20total%20carbon%20dioxide

Ranganathan, V. (1993). Electricity privatization the case of India. *Energy Policy*, 21(8), 875–880.

Schulze, M., Nehler, H., Ottosson, M., & Thollander, P. (2016). Energy management in industry–a systematic review of previous findings and an integrative conceptual framework. *Journal of Cleaner Production*, 112, 3692–3708.

Webb, M. G. (1985). Energy policy and the privatization of the UK energy industries. *Energy Policy*, 13(1), 27–36.

Wiedmann, T., & Minx, J. (2008). A definition of 'carbon footprint'. *Ecological Economics Research Trends*, 1, 1–11.

Wilberforce, T., Baroutaji, A., Soudan, B., Al-Alami, A. H., & Olabi, A. G. (2019). Outlook of carbon capture technology and challenges. *Science of the Total Environment*, 657, 56–72.

Wiser, R. H. (1998). Green power marketing: increasing customer demand for renewable energy. *Utilities Policy*, 7(2), 107–119.

3 The Sustainable Development Framework for Oil and Gas Resources

R. Suresh Kumar, S. Venkatesh, T. Ramakrishnan, and G. Gokilakrishnan

3.1 INTRODUCTION

Innovation within the upstream oil and gas sectors arises from a synthesis of factors and contributions, spanning across its value network. Many major oil and gas companies have their own research and development (R&D) centers dedicated to developing new technologies and methodologies for exploration, drilling, production, and refining processes. These centers often collaborate with academic institutions and technology partners to leverage expertise and resources. Companies specializing in oilfield services are involving in innovation by providing advanced technologies and services to oil and gas operators. These services include seismic imaging, drilling equipment, well construction, completions, and reservoir management. Service companies continuously invest in R&D for enhancing the efficiency, safety, and environmental performance. Apart from these companies, there are specialized technology providers focused on developing innovative solutions for specific challenges faced by the industry. These could be companies specializing in sensors, data analytics, robotics, artificial intelligence, or automation. Stricter regulations on emissions, waste management, and safety standards drive companies to invest in cutting-edge technologies mitigating environmental impact boosting operational efficiency. Multinational oil majors, such as ExxonMobil, Shell, BP, Chevron, and Total, invest significant resources in R&D and innovation. These companies have global operations and R&D centers in various countries, allowing them to leverage diverse talents and expertise. Countries with established technology hubs, such as Texas, California, Norway, and Canada, contribute extensively toward the innovation. These regions host numerous startups, research institutions, and technology accelerators focused on energy-related innovations. Emerging markets, particularly in regions like Africa, the Middle East, and Latin America, are increasingly investing in innovation to enhance their competitiveness in the global energy market. National oil companies (NOCs) in these regions are collaborating with international partners and investing in technology development to unlock their resource potential. Governments lead a

DOI: 10.1201/9781003452072-3

prominent role in fostering innovation through funding initiatives, tax incentives, and regulatory frameworks that support research and development. Countries with proactive government policies, such as Norway's focus on sustainable energy development or the United States' support for shale gas exploration, often notice accelerated innovation.

At present, the advancement of resources and the economic prospects are recognized as vital components for fostering national economic expansion. The anticipated prominence of resources in the global energy portfolio is poised to offset the dwindling output from mature fields post-2035, thus assuming a pivotal role in sustaining energy needs. Oil and gas production holds significance beyond ensuring energy security; it serves as a catalyst for the overall growth and development. However, there are exploration and production of hydrocarbons in the pose unique challenges and risks, including harsh environmental conditions, logistical complexities, and concerns about the impact on fragile ecosystems. As oil and gas explore their expansion, they face the imperative challenges to balance the economic benefits of resource development with the need to mitigate environmental risks and ensure sustainable practices. Addressing these challenges and fostering responsible, SD necessitates collaborative efforts from the government and the industry with environmental organizations.

3.1.1 Purpose of Study

This study endeavors to offer a comprehensive perspective aimed at enriching the understanding of the intricate interplay among oil and gas development, environmental preservation, socioeconomic prosperity, and innovation. By establishing a robust framework of indicators, policymakers, industry stakeholders, and civil society can make well-informed decisions and implement proactive measures promoting sustainable objectives in these sectors. This study provides view on the following:

- Offers a thorough and cohesive examination of SD principles and indicators that are customized to the unique context on the resources.
- Explores the nexus between strategic imperatives of SD and resource development, underscoring the interconnectedness of environmental, social, economic, and governance dimensions.
- Offers the rationale behind sustainability criteria within the environmental, socioeconomic, and innovative domains.
- Through retrospective analysis, forecasting, and quantitative indicator assessment, valuable insights into the dynamics of regional economic development are gleaned.
- Proposes recommendations aimed at fostering tailoring of a robust and actionable sustainable indicator for these resources.

Drawing from a survey meticulously crafted to explore these inquiries, this paper is poised to yield nuanced insights into the precise strategies, obstacles, and drivers of success linked with innovation across diverse segments of its value network and geographical locales. A noteworthy observation emerges that service firms demonstrate a propensity to file significantly more patents when juxtaposed with other organizational types. This suggests that service companies are particularly active in developing and protecting intellectual property related to innovative technologies and solutions in the industry. The survey findings underscore a notable 63% of deployed innovations traced back to service companies. This highlights the pivotal role that these providers assume in spearheading technological progress and introducing innovative solutions for effective utilization of these resources. Interestingly, neither universities nor research organizations either private or government has emerged as major contributors in providing newer insights and knowledge in the industry's research and development efforts. This suggests that the traditional sources of research and knowledge creation may not be as influential in driving innovation. These overarching themes and trends emphasize the pivotal role of service companies as primary catalysts for promoting innovation. Simultaneously, they shed light on the constrained contributions of traditional research institutions and reaffirm the enduring influence of the USA in shaping the industry's technological trajectory. Such insights serve as valuable inputs for strategic decision-making and resource allocation among companies and offer guidance for policymakers, aiming to bolster innovation and technological in these industries.

3.2 LITERATURE SURVEY

Major international oil companies (IOCs), such as ExxonMobil, Shell, BP, Chevron, and Total, possessed considerable resources and expertise devoted to driving technological advancements and enhancing operational efficiency. They invested heavily in exploring new reserves, developing advanced drilling techniques, enhancing recovery methods, and optimizing refining processes to maintain their competitive edge in the global energy market [1]. The landscape began to shift in the 1980s and 1990s, partly due to external factors, such as declining oil prices, increased competition from NOCs, and geopolitical uncertainties. As a result, many IOCs faced financial pressures and began to scale back their in-house R&D efforts, relying more on partnerships, outsourcing, and technology licensing arrangements to access innovative solutions developed by external vendors and service providers. This transition marked a significant change in the industry's R&D landscape, with a broader array of players contributing to technological innovation and knowledge creation. While the IOCs continued to play a prominent role in shaping the direction of R&D within the oil and gas sectors, their reduced investment in in-house programs led to a more diversified ecosystem where service companies, technology providers, research institutions, and governments all played increasingly important roles in driving innovation and advancing technology adoption within the industry. Technology is served as a cornerstone of the strategic priorities for IOCs, enabling them to achieve operational excellence, drive innovation, and adapt to changing market dynamics. However, the shift in the 1980s and 1990s toward reduced in-house R&D efforts reflected a broader reevaluation of investment priorities in response to evolving market conditions, competitive pressures, and financial considerations. Despite this shift, the legacy of technological innovation remains ingrained in the DNA of major IOCs, shaping their continued pursuit of excellence in the global energy landscape [2–4].

By maintaining comprehensive in-house R&D programs, IOCs aimed to foster a culture of innovation, build technical expertise, and secure proprietary technologies that could provide strategic advantages in the competitive landscape of the resources. These programs were held up by dedicated R&D facilities, multidisciplinary teams of scientists and engineers, and collaborations with academic institutions and research organizations [5]. The costs involved in modern-day R&D have in fact become increasingly higher [6,7]. The advancement of technology, coupled with the complexities of innovation, has led to escalating expenses across various stages of the R&D process, including fundamental research, prototyping, testing, and commercialization. In the past, significant breakthroughs across various industries often originated from in-house R&D teams within individual companies. However, contemporary researchers increasingly collaborate with external organizations to expand their access to diverse ideas and perspectives, fostering innovation across broader networks. By embracing external partnerships and networks, companies can enhance their innovation capabilities, expand their knowledge base, and remain competitive in an increasingly dynamic and interconnected world [8,9]. Numerous companies have adopted the open innovation concept [10] and embraced collaborative models for research and development [11–13]. This shift toward open innovation and collaborative R&D models reflects the recognition of the value of external inputs, diverse perspectives, and cross-industry knowledge exchange in driving innovation and addressing complex challenges. The incorporation of open innovation and collaborative research models underscores a strategic necessity to foster innovation, enhance competitiveness, and address complex challenges within a dynamic energy environment. Government agencies, vendors, oil companies, and universities now hold the potential to play significant roles in the research sector and development endeavors [14,15]. This shift toward a more diverse and collaborative R&D ecosystem reflected the recognition of the value of external expertise, cross-sector partnerships, and knowledge exchange in driving innovation and addressing industry challenges. The adoption of venture capital and alternative investment approaches reflects a strategic shift toward embracing external innovation, fostering entrepreneurship, and diversifying R&D investment portfolios within the industries. By leveraging the expertise, creativity, and agility of startups and innovators, companies can access novel technologies, accelerate innovation cycles, and drive sustainable growth in a rapidly evolving energy landscape [16–18]. This strategy of collaborating with competitors on research and development initiatives reflects the recognition of the mutual benefits and synergies that can be achieved through strategic partnerships, despite competitive dynamics within the industry. Collaborative innovation involved a diverse range of stakeholders working together to address industry challenges, drive technological advancement, and create value for stakeholders across the value chain. By leveraging their respective resources, expertise, and networks, these stakeholders collectively contribute to the industry's innovation ecosystem and ensure its long-term sustainability and competitiveness.

3.2.1 Radical and Product Innovation

Understanding the magnitude of change instigated by an innovation is essential for firms, policymakers, and industry stakeholders to anticipate market shifts, identify emerging opportunities, and align R&D investments with strategic priorities. While incremental innovations are crucial for maintaining competitiveness and fostering ongoing improvement, disruptive and radical innovations have the power to revolutionize industries, unlock fresh avenues for growth, and tackle pressing societal issues within and beyond the concern. In scholarly literature, certain technologies are often characterized based on diverse attributes and dimensions offering comprehensive understanding of their nature, impact, and implications.

a. The necessity of new skill sets, innovating companies proactively invest in talent development, training programs, and organizational capabilities to effectively harness the potential of emerging technologies and drive innovation-led growth. By strategically prioritizing skill acquisition, companies have been able to adeptly navigate shifting market dynamics, capitalizing promptly on emerging opportunities, and sustain a competitive advantage in a technology-centric environment [19].
b. By acknowledging the transformative potential of technologies that introduce new performance features, enhance existing ones, or reduce costs, companies can take proactive steps to invest in innovation, research, and development initiatives. By aligning these efforts with strategic objectives, businesses can stimulate growth and bolster competitiveness in dynamic and evolving markets [20].
c. By recognizing the transformative power of technologies that create entirely new lines of business, organizations proactively explore emerging opportunities, invest in innovation, and adapt their strategies to capitalize on the potential for growth and value creation in a rapidly changing world. By embracing disruptive technologies and cultivating a culture of innovation, companies position themselves to lead the industry, anticipate market shifts, and shape the future of business and society [21,22].
d. Incremental innovations support continuous improvement, sustaining competitiveness, and fostering the long-term growth within organizations. By prioritizing incremental innovation alongside more disruptive initiatives, companies balance risk and reward, maximize return on investment, and adapt to changing market conditions effectively [23].
e. Radical breakthrough innovations, such as 3D seismic mapping and horizontal drilling, mark significant milestones in this sector. They showcase the transformative capacity of technology and innovation to propel progress, prosperity, and sustainability in an ever-changing and interconnected global landscape. Through the adoption of disruptive technologies and the cultivation of an innovative culture, companies position themselves as leaders in shaping the future of energy and delivering value to stakeholders across the value chain [21,24,25].
f. By strategically integrating product and process innovations into their portfolios, companies can achieve synergies and maximize value creation and sustainability [20,26,27].
g. Distinguishing between product innovations and process innovations, organizations better understand the different dimensions of innovation and tailor their strategies and initiatives accordingly. While product innovations focus on creating value for customers through differentiated offerings, process innovations aim to enhance internal capabilities, operational performance, and organizational effectiveness [28].

3.3 MATERIALS AND METHODS

The essence of this study rests on the literature survey that consolidates details, spanning a wide array of topics, including current trends in the oil sector, the global energy domain, and fundamental aspects related to harnessing resource potential for its relevant projects. This review serves as a foundation for comprehending the current state of the industry, pinpointing emerging trends and its challenges contextualizing the unique dynamics of resource development within the sector. Utilizing this theoretical framework, this study seeks to address the sustainability of its activities, analyzes their effects on socioeconomic development, and provides insights to guide decision-making processes. Ultimately, the objective is to advocate for responsible and sustainable resource development, particularly in environmentally sensitive areas.

3.3.1 Methodological Approaches to Assessing Sustainability: Developing a System of Indicators

Indicative planning and assessment play crucial roles in the pursuit of SD objectives, providing a structured framework for guiding and evaluating progress toward sustainability goals. These components help to make the SD process more controllable and quantitatively measurable, moving it beyond abstract aspirations to concrete actions and outcomes. Indicators function as indispensable tools in this process, empowering stakeholders to measure, monitor, and analyze the progress toward sustainability targets. They facilitate the identification of areas requiring improvement and enable adjustments to the development trajectory as necessary. Creating a system of indicators for SD requires a collaborative and multi-stakeholder approach. By maintaining a balance between global aspirations, local contexts, and corporate goals, sustainability indicators can effectively guide decision-making, promote accountability, and advance progress toward a more sustainable and inclusive future. The importance of establishing a comprehensive system of SD indicators to enable informed decision-making across various management levels was underscored at the Earth Summit, held in 1992 in Rio de Janeiro. It represented a pivotal moment in the collective global environmental movement to tackle urgent environmental and developmental challenges confronting the planet. In fact, a diverse array of SD criteria and indicators was developed to suit systems of different scales, ranging from global to local levels. These indicators are developed to assess the progress toward sustainability goals and inform policy-making and decision-making processes [29–32]. An exemplary indicator is the global indicator developed for the assessment of SD goals (SDGs). This framework, known as the SDG Indicators, was developed by the United Nations Statistical Commission in collaboration with various stakeholders and serves as a key tool for tracking progress toward achieving the SDGs by the year 2030 [33]. Wu [34] developed a rationale for crafting quantitative indicators to evaluate SD, aiming to harmonize the complex interplay among the different facets of SD. The author argued that by quantifying key aspects of sustainability, it becomes feasible to manage and navigate the complexities inherent in SD initiatives effectively. The Balanced Scorecard, introduced by Robert and colleagues [35], provided a framework for SD indicators. This framework offers a structured approach to aligning sustainability goals with organizational strategies and priorities. This model is widely embraced across due to its ability to encapsulate the interconnectedness of causal relationships and strategic priorities.

3.3.2 SD Indicators

Development of SD indicators involves the following methods:

1. **Comprehensive** Assessment **of** Sustainability

   It considers the interconnectedness of these dimensions, namely, economy, society, and environment for sustainability. This approach involves integrating various sets of indicators to highlight their individual importance within the overall system. Weights are commonly assigned to each indicator to represent its relative significance or contribution to the overall sustainability assessment. However, the process of assigning weights often relies on expert judgments, which may introduce subjectivity and bias into the weighting process. As a result, the required degree of objectivity may not always be achieved. Experts may have different perspectives, values, and priorities, leading to subjective judgments about the importance of different indicators. This subjectivity can result in varying weight assignments, depending on the expertise and opinions of the individuals involved. The process of assigning weights based on expert assessments may lack transparency, making it difficult for stakeholders to understand how weights were determined and to assess the validity of the weighting decisions. Experts may struggle to accurately assess the relative importance of indicators in such complex systems, leading to challenges in assigning meaningful weights.
2. **Designing a System of Indicators**

   It involves meticulously selecting and organizing indicators to capture the distinct facets of SD, thereby reflecting separate aspects of sustainability. The second method of integrating indicators for SD in fact presents challenges in identifying the relationships between indicators and assessing their relative importance or contribution to the overall system. This method typically involves aggregating indicators without explicitly considering their interdependencies or weights. This method can in fact present challenges when attempting to compare the levels of sustainability among individual objects, projects, or systems.

3.3.3 Global 100 Index

The Global 100 list, compiled and published annually by Corporate Knights, a Canadian media and research company, ranks the most prominent indexing. This index evaluates thousands of publicly traded companies strictly on various environmental, social, and governance (ESG) indicators. It has four groups as listed below:

1. Financial Management

   It evaluates how effectively companies manage their financial resources while considering sustainability factors. Key indicators may include financial stability, profitability, revenue growth, return on investment (ROI), and debt management. Companies with strong financial management practices are better positioned to achieve long-term sustainability goals and deliver value to shareholders [36].

2. Resource Management

   It assesses utilization of natural resources and their efforts to minimize environmental impacts. Companies that demonstrate responsible resource management practices are more likely to mitigate environmental risks and contribute to SD [36].

3. Employee Management

   It focuses on how companies manage their human capital and promote employee well-being. Indicators may include employee diversity and inclusion, workplace safety, labor practices, employee training and development, and employee satisfaction. Companies that prioritize employee management are better positioned to attract and retain talent, foster innovation, and enhance productivity.

4. Deductions Due to Sanctions

   It reflects deductions applied to companies' overall scores due to involvement in sanctions, legal violations, or unethical practices. Companies may receive deductions for activities, such as environmental violations, human rights abuses, corruption, or violations of international labor standards. Deductions serve as a mechanism to hold companies accountable for noncompliance with ethical and legal standards and promote responsible corporate behavior.

3.3.4 SAM Corporate Sustainability Assessment (CSA)

An annual evaluation was conducted by SAM, a division of S&P Global, to assess the sustainability performance of an industry. SAM's CSA is one of the most comprehensive assessments of corporate sustainability practices and is widely recognized by investors, stakeholders, and companies alike. The methodology employed in the SAM Corporate Sustainability Assessment redefined the Dow Jones Sustainability Index family (DJSI), which stands as one of the most renowned sustainability indices worldwide. The DJSI evaluates companies reflecting the three pillars of sustainability based on environmental, economic, and social performance. The DJSI promotes a holistic understanding of sustainability and encourages companies to adopt more responsible business practices [37].

3.3.5 Global Reporting Initiative (GRI)

GRI, established in 1997, is an independent international organization renowned for crafting a widely adopted framework for sustainability reporting. This framework offers guidelines and standards enabling organizations to transparently and comparably report their economic, ESG performance. GRI standards cover an extensive array of sustainability topics, encompassing governance, anti-corruption measures, human rights, labor practices, environmental impacts, and product responsibility. Designed to cater to organizations of all sizes, sectors, and geographic locations, these standards are utilized by numerous entities worldwide, including corporations, governments, nonprofit organizations, and academic institutions, thereby establishing GRI reporting as a universally recognized and embraced practice [38].

3.3.6 Development of Oil and Gas Resources: Sustainability Assessment

The interest in Arctic oil and gas development by the global scientific community reflects a complex interplay of economic, technological, geopolitical, and environmental factors. While there are opportunities for energy security and economic development, there are also risks and challenges to be carefully considered and managed to ensure the sustainability of Arctic ecosystems and communities [39]. The emergence of trends toward environmental friendliness and sustainability has shifted the focus of oil and gas projects beyond purely economic considerations. In today's landscape, successful project implementation necessitates comprehensive assessments considering economic outcomes and social and environmental ramifications. S. Kirsch's [40] conclusion highlighted the importance of considering the limitations and challenges associated with nonrenewable resource extraction and the need for broader discussions and actions toward achieving a more sustainable and equitable global energy and resource system. Kristoffersen [41] argument underscored the need for careful consideration of environmental concerns in decision-making processes related to Arctic resource development. It emphasizes the significance of embracing a proactive stance, honoring indigenous wisdom and rights and placing emphasis on preserving and cultivating sustainable practices, all aimed at safeguarding the enduring vitality and adaptability of Arctic ecosystems and communities. The assumption made by the authors [42] reflected a nuanced understanding of sustainability that recognizes the complexity of balancing competing interests and objectives in natural resource management. By emphasizing the importance of considering long-term benefits and impacts across environmental, social, and economic dimensions, their approach seeks to promote more responsible and sustainable practices in resource development. Though sustainability entails economic, social, and environmental aspects, the emphasis on environmental factors regarding oil and gas development underscores the acknowledgment of the distinctive ecological hurdles and hazards linked with resource extraction in the region. However, achieving true sustainability required a comprehensive and integrated approach that addresses all dimensions of development while safeguarding the long-term health and resilience of Arctic ecosystems and communities [43,44]. The actions taken by UBS, Wells Fargo, Goldman Sachs, and other banks to restrict investments reflected a broader trend toward sustainable finance and climate-conscious investing. These actions signal the recognition of the financial, environmental, and reputational risks associated with fossil fuel investments and a commitment toward sustainable and low-carbon economy [45].

3.4 ISSUES ASSOCIATED WITH THE SD OF OIL AND GAS RESOURCES

The SD concept mirrors an evolving acknowledgment of the necessity to harmonize economic objectives with environmental preservation and societal accountability. While the industry's initial focus was on addressing environmental safety in resource exploitation, SD principles have since evolved to encompass broader considerations of social equity, economic viability, and long-term environmental stewardship [46]. While the concept of SD was in fact reflected in various UN documents and initiatives prior to 2015, the formal adoption of the SDGs and their inclusion in the 2030 Agenda will mark a significant milestone in global efforts to advance SD on a global scale [43,47]. The transition toward qualitative development and the incorporation of SDGs into business practices signify a commendable progression toward enhanced sustainability and conscientious management of natural resources. By embracing the SDGs and adopting new practices, companies can contribute to addressing global challenges while ensuring long-term value creation and resilience in a rapidly changing world [47].

The definition provided by Nurtdinov and team [48] underscored the transformative nature of SD and its broader implications for society, economy, and the environment. It underscores the imperative for a comprehensive and unified strategy to development that nurtures harmony, equilibrium, and resilience amid multifaceted and interlinked challenges. This complexity stems from the varied and interrelated effects on ecosystems, communities, and economies. Various factors, including resource depletion, climate change, environmental degradation, social conflicts, and economic volatility, collectively contribute to the multifaceted nature of SD within the sector. Since SD is inherently complex, there's often no unanimous agreement on its essence or the means to achieve it. Different stakeholders may prioritize various sustainability aspects, resulting in divergent perspectives and approaches. This lack of consensus can pose challenges in developing cohesive strategies and policies for SD within the industry. The energy sector, including oil and gas, is subject to global instability driven by factors, such as geopolitical tensions, fluctuating commodity prices, technological advancements, regulatory changes, and evolving consumer preferences. This volatility can impact the sustainability, affecting investment decisions, project viability, and long-term planning for SD.

3.5 ENVIRONMENTAL SUSTAINABILITY OF THE DEVELOPMENT OF OIL AND GAS RESOURCES

As global focus intensifies on mitigating climate change and transitioning to a low-carbon economy, there's a noticeable trend toward adopting carbon-neutral practices in the energy sector. Exploitation of oil and gas fields presents both challenges and opportunities in this regard as it requires balancing energy demand with environmental sustainability and carbon emission reduction goals. Several oil and gas producing nations, notably those with extensive Arctic territories, have implemented national environmental policies and regulations aimed at safeguarding the delicate ecosystems of the Arctic. These policies are geared toward mitigating the environmental repercussions of resource extraction and fostering SD in the region. By prioritizing environmental concerns, oil and gas producing countries align with their national interests, fulfill international obligations, enhance corporate reputations, meet stakeholder expectations, and advance the long-term sustainability of energy projects. It has become increasingly vital to consider the immediate impacts and its long-term consequences on ecosystems, communities, and the global climate. This holistic approach ensures that every project executed need to be aligned with the

SD goals while minimizing negative environmental outcomes. The following key aspects should be considered:

a. Resource Extraction Techniques
   Evaluate the environmental impact of extraction techniques, such as hydraulic fracturing (fracking), offshore drilling, and tar sands extraction. Each method has unique environmental risks, including water contamination, habitat destruction, and greenhouse gas emissions [49].
b. Water Usage and Pollution
   Assess the water usage and potential for water pollution associated with oil and gas operations. This includes evaluating the sourcing of water for extraction processes, the management of produced water (contaminated water extracted along with oil and gas), and the risks of spills or leaks contaminating water sources [49].
c. Air Emissions
   Examining greenhouse gases, along with air pollutants, such as volatile organic compounds and nitrogen oxides, stemming from oil and gas operations is crucial. Let's assess strategies to diminish these emissions through advancements in technology, adherence to regulations, and embracing cleaner energy alternatives [50].
d. Waste Management
   Explore the handling of waste produced covering drilling waste, produced water, and solid waste. Evaluate approaches for waste management, recycling, disposal, and remediation to mitigate environmental effects and meet regulatory standards [51].
e. Biodiversity Conservation
   Evaluate the impact of process development on local ecosystems and biodiversity. Consider measures to protect sensitive habitats, mitigate habitat fragmentation, and minimize the disturbance of wildlife during exploration, extraction, and transportation activities [51].
f. Land Use and Habitat Preservation
   Assess the land use changes and habitat destruction associated with oil and gas development, including deforestation, wetland degradation, and urbanization. Consider strategies for minimizing land disturbance, restoring impacted habitats, and preserving critical ecosystems [52].
g. Climate Change Mitigation
   Consider the contribution of such process development to climate change through the release of greenhouse gas emissions. Evaluate measures to minimize emissions, improve energy efficiency, and transition to cleaner energy sources to mitigate climate impacts and support global climate goals [53].
h. Community Health and Safety
   Evaluate the potential impacts of oil and gas development on community health and safety, including air and water quality, noise pollution, and the risk of accidents or spills. Consider measures to protect public health, ensure worker safety, and engage with local communities to address concerns and mitigate risks [54].

i. Regulatory Compliance and Enforcement

Assess the effectiveness of regulatory frameworks and enforcement mechanisms for overseeing oil and gas operations and ensuring environmental compliance. Consider opportunities to strengthen regulations, improve monitoring and enforcement, and promote transparency and accountability [54].

The adoption of electrification, hydrogen production, and low-carbon fuels has improvised the potential to mitigate environmental impact and also enhance operational efficiency, bolster energy security, and capitalize on emerging market dynamics. Embracing innovation, investing in sustainable technologies, and fostering collaboration across the energy sector will be essential for unlocking the full potential of these opportunities and driving the industry toward a more sustainable future [55]. The claim that oil and gas resource development can align with climate objectives through transformation is substantiated by evidence demonstrating a reduction in greenhouse gas emissions despite escalating production levels.

3.6 DISCUSSION

In addition to concerns about the financial viability of ambitious projects amid energy market volatility and declining demand and prices, particularly for oil, there is a pressing need to address the environmental and social implications. Achieving a balance between harnessing resource potential, preserving the natural environment, and maintaining socioeconomic stability presents a significant challenge globally. Ensuring the sustainability demands a multifaceted approach: meeting domestic hydrocarbon needs while maintaining stable exports, mitigating environmental effects, fostering innovative technologies and infrastructure, optimizing production capabilities, improving transportation and community facilities, and creating new avenues for high-tech employment. Currently, significant environmental risks persist, particularly in the continental shelf and the advancement of trans-Arctic shipping routes. Leakage of gas and spilling of oils during transit represent significant threats to the delicate northern marine ecosystems. Moreover, the substantial extraction of hydrocarbons exacerbates global warming. Addressing these challenges outweighs the pursuit of economic and geopolitical gains. A comprehensive evaluation of production sustainability can be achieved by incorporating indicators related to competency development, initiatives aimed at protecting the habitats and culture of indigenous communities, and advancements in social and transportation infrastructure within the region. Additionally, innovative sustainability metrics may include advancements in production techniques and the utilization of digital techniques to address production and managerial challenges.

3.7 CONCLUSION

The incorporation of SD in the energy sector globally is reshaping the perspective of governance in the field of oil and gas sectors. This integration is prompting a reorientation of priorities within the sector. Three key priorities have emerged as focal points as briefed below:

a. Environmentally Friendly Production and Transportation

There is a growing emphasis on minimizing the impact on the environment from exploration to extraction and transportation. This entails adopting cleaner technologies, implementing rigorous environmental safeguards, and mitigating risks associated with spills, leaks, and emissions. Companies are progressively allocating resources to research and development aimed at bolstering energy efficiency, curbing carbon emissions, and exploring alternative energy sources. Moreover, there is a concentrated endeavor to ensure that transportation methods, whether via pipeline, ship, or other means, prioritize safety and environmental preservation.

b. Social Responsibility and Socioeconomic Development

Companies are acknowledging the importance of nurturing positive relationships with local communities and indigenous populations in the regions where they operate. This entails engaging in meaningful dialogue, respecting cultural heritage and land rights, and actively participating in the socioeconomic advancement of these areas. Companies recognize the significance of investment in local infrastructure, education, healthcare, and job training programs to bolster community well-being and promote economic empowerment. Additionally, there is an increasing focus on fostering diversity, equity, and inclusion within the workforce, as well as supporting initiatives that strengthen social cohesion and resilience.

c. Economic Efficiency of Projects

Economic viability remains a critical factor in the development of these projects. Companies are seeking innovative ways to improve its operational efficiency, costs incurred, and optimize the utilization of resource without compromising sustainability goals. This may involve leveraging technology advancements, streamlining processes, and pursuing partnerships and collaborations to maximize returns on investment. Additionally, companies are increasingly factoring in long-term financial risks associated with climate change, regulatory compliance, and market volatility when evaluating project feasibility and profitability.

ACKNOWLEDGMENT

This project is funded by the DST-FIST, Government of India, Ref. No: SR/FST/College-/2022/1300.

REFERENCES

1. Economides, M. (2000). *The Color of Oil: The History, the Money and the Politics of the World's Biggest Business.* Round Oak Publishing Co, Katy, TX.
2. Wilkins, M. (1975). The oil companies in perspective. *Daedalus*, 104, 159–178.
3. Howarth, S., & Jonker, J. (2007). *A History of Royal Dutch Shell. Vol. 2: Powering the Hydrocarbon Revolution, 1939–1973.* Oxford University Press, New York.
4. Priest, T. (2009). *The Offshore Imperative: Shell Oil's Search for Petroleum in Postwar America.* Texas A&M University Press, College Station, TX.

5. Sharma, A. K. (2005). *Technology Collaboration between Competitive Firms in the Upstream Oil Industry: The Effect of Operational Control on Collaborative Behavior.* Case Western Reserve University, Cleveland, OH.
6. Kumpe, T., & Bolwijn, P. (1988). Manufacturing: the new case for vertical integration. *Harvard Business Review*, 66(2), 75–81.
7. Mayers, A. J. C., & Brenner, Y. S. (1995). Make or buy: the potential subversion of corporate strategy–the case of Philips. *International Journal of Social Economics*, 22(4), 4–11.
8. Quinn, J. B., & Hilmer, F. G. (1994). Strategic outsourcing. *MIT Sloan Management Review*, 35(4), 43.
9. Rigby, D., & Zook, C. (2002). Open-market innovation. *Harvard Business Review*, 80, 5–12.
10. Chesbrough, H. W. (2006). The era of open innovation. *Managing Innovation and Change*, 127(3), 34–41.
11. Chesbrough, H. W. (2003). *Open Innovation: The New Imperative for Creating and Profiting from Technology.* Harvard Business Press, Brighton, MA.
12. Verloop, J. (2006). The Shell way to innovate. *International Journal of Technology Management*, 34(3–4), 243–259.
13. Ramirez, R., Roodhart, L., & Manders, W. (2011). How Shell's domains link innovation and strategy. *Long Range Planning*, 44(4), 250–270.
14. Dennis, R., Jones, T., & Roodhart, L. (2011). Technology foresight: the evolution of the Shell gamechanger technology futures program. In S.P. MacGregor & T. Carleton (ed.) *Sustaining Innovation: Collaboration Models for a Complex World* (pp. 153–165). Springer, New York.
15. Acha, V. L. (2002). Framing the past and future: the development and deployment of technological capabilities by the oil majors in the upstream petroleum industry. Available at SSRN: https://ssrn.com/abstract=1357624 or http://doi.org/10.2139/ssrn.1357624.
16. Acha, V., & Cusmano, L. (2005). Governance and co-ordination of distributed innovation processes: patterns of R&D co-operation in the upstream petroleum industry. *Economics of Innovation and New Technology*, 14(1–2), 1–21.
17. Hansen, M. T., & Birkinshaw, J. (2007). The innovation value chain. *Harvard Business Review*, 85(6), 121–30.
18. Shah, C. M., Zegveld, M. A., & Roodhart, L. (2008). Designing ventures that work. *Research-Technology Management*, 51(2), 17–25.
19. Crump, J. G. (1997). Strategic alliances fit pattern of industry innovation. *Oil and Gas Journal*, 95(13), 59–63.
20. Allan, A. (1998). *Innovation Management: Strategies, Implementation and Profits.* Oxford University Press, New York.
21. Leifer, R. (2000). *Radical Innovation: How Mature Companies Can Outsmart Upstarts.* Harvard Business Press, Brighton, MA.
22. Bozdogan, K., Deyst, J., Hoult, D., & Lucas, M. (1998). Architectural innovation in product development through early supplier integration. *R&D Management*, 28(3), 163–173.
23. McDermott, C. M. (1999). Managing radical product development in large manufacturing firms: a longitudinal study. *Journal of Operations Management*, 17(6), 631–644.
24. Gilbert, C. (2003). The disruption opportunity. *Mit Sloan Management Review*, 44(4), 27–32.
25. Managi, S., Opaluch, J. J., Jin, D., & Grigalunas, T. A. (2005). Technological change and petroleum exploration in the Gulf of Mexico. *Energy Policy*, 33(5), 619–632.
26. Martin, J. M. (1996). Energy technologies: systemic aspects, technological trajectories, and institutional frameworks. *Technological Forecasting and Social Change*, 53(1), 81–95.

27. Yergin, D. (2012). *The Quest: Energy, Security, and the Remaking of the Modern World*. Penguin Book, New York.
28. Tidd, J., Bessant, J., Pavitt, K., 2001. *Managing Innovation: Integrating Technological, Market, and Organizational Change*, 2nd ed. John Wiley & Sons Ltd, New York.
29. Burgelman, R. A., Maidique, M. A., & Wheelwright, S. C. (1996). *Strategic Management of Technology and Innovation* (Vol. 2, p. 37). Irwin, Chicago, IL. https://isolaralliance.org/about/background
30. Schilling, M. A. (2017). *Strategic Management of Technological Innovation*. McGraw-Hill, New York.
31. United Nations. The System of Environmental-Economic Accounting (SEEA). Available online: https://seea.un.org/ru (accessed on 3 September 2021).
32. United Nations. Human Development Index (HDI). Available online: https://hdr.undp.org/en/content/human-developmentindex-hdi (accessed on 3 September 2021).
33. The World Bank. Indicators. Available online: https://data.worldbank.org/indicator (accessed on 3 September 2021).
34. FDI, OECD. Qualities Indicators: Measuring the Sustainable Development Impacts of Investment. Available online: https://www.oecd.org/fr/investissement/fdi-qualities-indicators.htm (accessed on 3 September 2021).
35. SDG Indicators. Global Indicator Framework for the Sustainable Development Goals and Targets of the 2030 Agenda for Sustainable Development. Available online: https://unstats.un.org/sdgs/indicators/indicators-list/ (accessed on 3 September 2021).
36. Wu, J., & Wu, T. (2012). Sustainability indicators and indices: an overview. *Handbook of Sustainability Management*, 2012, 65–86.
37. Menshchikova, V. I., & Sinopolets, N. V. (2011). System of indicators for assessing the sustainable development of the region's economy. *Social Economics Phenomenon Process*, 5–6, 155–160.
38. The 2019 Global 100: Overview of Corporate Knights Rating Methodology. Available online: https://www.corporateknights.com/wp-content/uploads/2018/10/2019-Global-100_Methodology-Final.pdf?v=20181205
39. Measuring Intangibles ROBECOSAM's Corporate Sustainability Assessment. Available online: https://www.spglobal.com/spdji/en/documents/additional-material/robeco-sam-measuring-intangibles.pdf
40. Global Sustainability Standards Board. Available online: https://www.globalreporting.org/standards/global-sustainabilitystandards-board
41. Coronacrisis: The Impact of COVID-19 on the Fuel and Energy Sector in the world and in Russia.2020. Available online: https://energy.skolkovo.ru/downloads/documents/SEneC/Research/SKOLKOVO_EneC_COVID19_and_Energy_sector_RU.pdf
42. Kirsch, S. (2010). Sustainable mining. *Dialectical Anthropology*, 34, 87–93.
43. Kristoffersen, B., & Langhelle, O. (2017). Sustainable development as a global-Arctic matter: imaginaries and controversies. In K. Keil & S. Knecht (ed.) *Governing Arctic Change* (pp. 21–41). Palgrave Macmillan, London.
44. Amezaga, J. M., Rötting, T. S., Younger, P. L., Nairn, R. W., Noles, A. J., Oyarzún, R., & Quintanilla, J. (2011). A rich vein? Mining and the pursuit of sustainability. *Environmental Science and Technology*, 1, 21–26.
45. Stipo, F., Thorhaug, A., Jackson, R., Butler, K., Gibbs, R., Gray, J., & Zak, B. (2012). The future of the arctic: a key to global sustainability. *Cadmus*, 1(5), 42–52.
46. The 2030 Decarbonization Challenge. The Path to the Future of Energy. Deloitte 2020. Available online: https://www2.deloitte.com/content/dam/Deloitte/global/Documents/Energy-and-Resources/gx-eri-decarbonization-report.pdf.
47. Assembly, U. G. (2015). Resolution adopted by the General Assembly on 25 September. Transforming our world: the 2030 Agenda for Sustainable Development. https://www.un.org/en/development/desa/population/migration/generalassembly/docs/globalcompact/A_RES_70_1_E.pdf.

48. United Nations. Paris Agreement. 2015. Available online: https:////unfccc.int/files/essential_background/convention/application/pdf/english_paris_agreement.pdf.
49. UNDP, I. (2017). Mapping the oil and gas industry to the sustainable development goals: an atlas. *Sustainable Development Goals.* Available from: https://www.undp.orgecontentedameundpelibraryeSustainable%20DevelopmenteExtractveseFor%20CommentNMapping%20the%20Oil%20and%20Gas%20industry%20to%20the%20Sustainable%20Development%20Goals.
50. Nurtdinov, R. M., & Nurtdinov, A. R. (2012). From the theory of economic growth to the concept of sustainable development: issues of rethinking. *Vestnik Kazanskogo Tehnologicheskogo Universiteta*, 23, 136–141.
51. Bobylev, S. N., Kiryushin, P. A., & Kudryavtseva, O. V. (2019). *Green Economy and Sustainable Development Goals for Russia: Collective Monograph.* Faculty of Economics of Lomonosov Moscow State University, Moscow, Russia, 284.
52. Peters, G. P., Nilssen, T. B., Lindholt, L., Eide, M. S., Glomsrød, S., Eide, L. I., & Fuglestvedt, J. S. (2011). Future emissions from shipping and petroleum activities in the Arctic. *Atmospheric Chemistry and Physics*, 11(11), 5305–5320.
53. McGlade, C., & Ekins, P. (2015). The geographical distribution of fossil fuels unused when limiting global warming to 2°C. *Nature*, 517(7533), 187–190.
54. Zhavoronkova, N. G., & Agafonov, V. B. (2019). Strategic directions of legal provision of environmental safety in the Arctic zone of the Russian Federation. *Actual Problems of Russian Law Journal*, 7, 161–171.
55. Kapoor, A., Fraser, G. S., & Carter, A. (2021). Marine conservation versus offshore oil and gas extraction: Reconciling an intensifying dilemma in Atlantic Canada. *The Extractive Industries and Society*, 8(4), 100978.

4 Digital Twin Technologies in Manufacturing Industries

L. Venkatesh, M. Sowrirajan, M. Arulraj, V. Rajkumar, and R. Manikandan

4.1 INTRODUCTION

Digital twin (DT) technology is a transformative approach that generates a virtual representation of physical entities, facilitating improved monitoring, prediction, and optimization of their performance. This technology is particularly significant with reference to Industry 4.0, where it integrates sophisticated manufacturing techniques with real-time data to improve processes and drive innovation. A dynamic digital representation faithfully replicates a tangible object, process, or mechanism, constantly updating through real-time data gathered from sensors and other sources, ensuring dependable simulations. This capability facilitates predictive analytics, enabling organizations to monitor performance and optimize operations effectively (Javaid et al., 2023). Michael Grieves introduced DT in his 2003 Product Lifecycle Management (PLM) course. The DT framework included three conceptual dimensions: a physical object, a virtual counterpart, and a connection. However, in 2010, National Aeronautics and Space Administration (NASA) released a definition of DT as "an integrated multiphysics, multiscale simulation of a vehicle or system that uses the best available physical models, sensor updates, fleet history, etc., to mirror the life of its corresponding flying twin" (Tao et al., 2019). The concept of DT originated in the aerospace and automotive industries but has since expanded to various sectors, including healthcare, energy, and consumer goods. As industries face increasing pressure to enhance productivity and reduce costs, DT offers a solution that leverages data and analytics to drive continuous improvement. Research indicates that the adoption of DT technology can lead to advantages including faster time-to-market, better product quality, and enhanced operational efficiency (Kritzinger et al., 2018). It comprises the physical entity (PE), the digital model, and data connections. The PE refers to the actual object or system being replicated, serving as the foundation for DT. The digital model is a virtual representation that simulates the properties and behaviors of the PE, allowing for real-time monitoring and analysis. Finally, data connections involve continuous real-time data streams that synchronize the physical and digital worlds, ensuring that DT remains accurate and relevant (Grieves, 2014). This integration of components enables organizations to leverage DT for enhanced

DOI: 10.1201/9781003452072-4

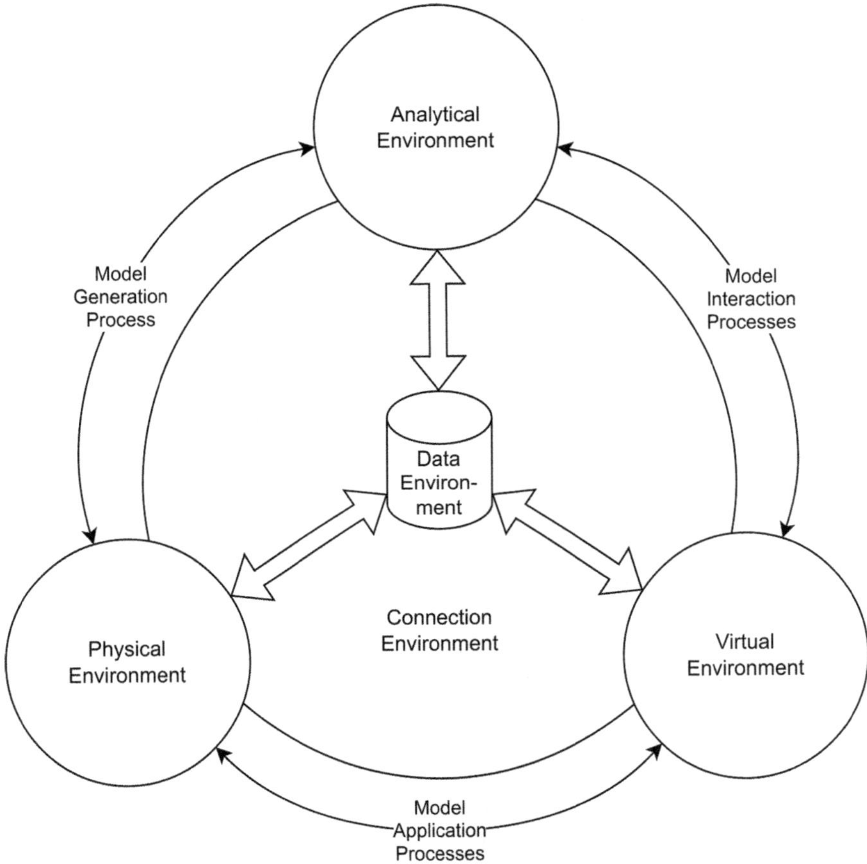

FIGURE 4.1 Components of a DT (Grübel et al., 2022).

decision-making and operational efficiency in various applications, encompassing production and maintenance procedures (Grübel et al., 2022) (Figure 4.1).

DT technology distinguishes itself from conventional simulation models by its ability to provide real-time reflection, interaction, convergence, and self-evolution. DTs exhibit strong synchronization with their physical counterparts, guaranteeing precise and immediate updates and monitoring. They provide the complete integration of system phases, components, and services, allowing for easy access to data and optimization. By integrating historical data with up-to-the-minute information, DTs provide a comprehensive comprehension of system behavior and performance. The reciprocal link between real and virtual worlds guarantees prompt reflection of changes, enabling ongoing feedback and improvement. In addition, DTs can update their data in real time, continuously improving their virtual models by integrating fresh data and insights. This process enhances the accuracy and prediction capacities of the DTs. DT technology has these attributes, making it a potent instrument for attaining enhanced efficiency, dependability, and ingenuity across many sectors. DT has distinct properties that set it apart from conventional simulation models (Onaji et al., 2022).

4.2 APPLICATIONS OF DIGITAL TWIN (DT)

DT technology finds applications across various stages of the product life cycle, enhancing processes in product design, manufacturing, service and maintenance, supply chain management, healthcare, and urban planning (Attaran & Celik, 2023; Lim et al., 2020; Tao et al., 2019).

a. **Product Design:** DTs streamline product design via digital modeling and simulation. Designers may visualize physical factors and simulate product performance using virtual models. Early design problem discovery promotes flexibility. Simulations using digital models enable configuration testing without physical prototypes, saving time and money and eliminating design mistakes and delays. Designers may get input and make educated choices from suppliers, consumers, and regulatory agencies by sharing digital models and simulations. This ensures the final product satisfies stakeholder demands.
b. **Manufacturing:** DTs provide real-time manufacturing process monitoring, optimization, and control. Continuously updating the digital model with PE data gives manufacturers system performance and condition insights. This allows problem discovery, process optimization, and efficiency gains. Sensor data allows DTs to forecast equipment failures and offer proactive maintenance, decreasing downtime and maintenance costs and boosting reliability and performance. Digital model simulations uncover bottlenecks and inefficiencies, improving production, quality, and cost.
c. **Service/Maintenance:** DTs offer efficient customer service and maintenance. Digital models allow manufacturers to monitor product performance and condition in real time and provide proactive maintenance and support. Customer satisfaction and loyalty increase as downtime and maintenance costs decrease. DTs allow service professionals to diagnose and fix faults remotely using the digital model. Time and money are saved, and travel delays are reduced. Additionally, DTs provide proactive and personalized help. Analyzing digital model data identifies faults and recommends targeted maintenance, enhancing product and service dependability and performance.
d. **Supply Chain Management:** DTs improve products, information, and financial flow. Digitalizing supply chain networks improves decision-making and efficiency. DTs give accurate real-time demand projections based on sales, market trends, and external variables, allowing organizations to modify production and inventory levels to avoid stockouts and overstock. Companies may find the best cost-effective and environmentally friendly logistics solutions by simulating multiple transportation routes and modalities. DTs also improve supply chain resilience by offering real-time interruption visibility. Companies may develop contingency plans and guarantee operational continuity by monitoring the supply chain network for early anomaly identification and predictive effect assessment.

e. **Healthcare:** Patient outcomes and treatment are being transformed by DT technology. Digital patient replicas enable doctors to simulate and analyze treatment choices for personalized and accurate care. DTs use electronic health records, genetic data, and real-time monitoring equipment to assess a patient's health. This permits customized therapies, enhancing effectiveness and decreasing side effects. DTs improve surgical planning and training. Surgeons practice and improve abilities via virtual surgical simulations, enhancing patient safety. Healthcare practitioners may remotely monitor patients, recognize early health decline, and respond using real-time information.

f. **Urban Planning:** DT is used in urban planning to make cities smarter and more sustainable. Simulations and analyses of urban settings using digital replicas enable informed decision-making and effective resource management. In infrastructure management, DTs monitor important infrastructure's condition and performance to identify maintenance requirements and prioritize repairs for safety and durability. DTs simulate traffic flow and congestion to improve traffic management. Designing effective transit routes and improving traffic flow reduces congestion. By modeling planning choices' environmental consequences, DTs promote sustainable urban development. Energy, air quality, and waste management data help develop eco-friendly cities.

4.3 DIGITAL TWIN (DT) TECHNOLOGY IN MANUFACTURING

The manufacturing sector is now experiencing a substantial revolution propelled by the incorporation of cutting-edge technologies such as cyber-physical systems (CPS), the Internet of Things (IoT), and artificial intelligence (AI). Industry 4.0 is a progressive approach that strives to enhance adaptability, facilitate quick design changes, and improve the flexibility of staff training in industrial processes (Attaran & Celik, 2023). As the industry embraces these technological advancements, the idea of being environmentally friendly is becoming more significant. Manufacturers must now consider not only productivity and efficiency but also the environmental impact and resource optimization of their operations. These advanced virtual models replicate physical assets, operations, and systems, enabling real-time tracking, simulation, and refinement. This dynamic digital counterpart helps industries optimize energy usage and manage emissions, ushering in a new era of sustainable and efficient manufacturing practices (Qi & Tao, 2018).

DT technology is really vital in this transformation, bridging the gap between physical and virtual worlds to create a seamless flow of data and insights (Fuller et al., 2020). By developing a real-time digital reproduction of a PE, DT enables manufacturers to monitor, simulate, and optimize their systems in ways previously unimaginable (Glaessgen & Stargel, 2012). DT technology finds its applications in different stages of the product life cycle, ranging from design and manufacturing to service and maintenance (Qi & Tao, 2018). It also explores the enabling technologies that underpin DT-driven sustainable intelligent manufacturing, such as big data analytics, AI, and the IoT. Data from many sources, including sensors, IoT devices, and historical documents, is integrated by DTs to create a comprehensive and continuously updated digital model. This integration allows for precise tracking of performance, detection of anomalies, and preventive maintenance, which reduces unplanned downtime and boosts operational efficiency generally (Grieves & Vickers, 2016). By leveraging advanced analytics and machine learning algorithms, DT can simulate different scenarios and predict outcomes, enabling manufacturers to make informed decisions that optimize energy consumption and reduce emissions (Tao et al., 2019). The application of DT technologies in manufacturing extends to various facets of the industry. In the automotive sector, DT is used to simulate production lines, identify bottlenecks, and optimize workflows, resulting in significant energy savings (Uhlemann et al., 2017). In the chemical industry, these technologies help monitor complex chemical processes, ensuring optimal operating conditions and reducing energy waste. Furthermore, in electronics manufacturing, DT assists in designing energy-efficient production systems, minimizing resource consumption, and reducing the carbon footprint (Kritzinger et al., 2018).

DT technologies have a significant advantage in their capacity to enable and support proactive rather than reactive management of industrial processes. Through the provision of up-to-the-minute information and the ability to forecast future events, DT empowers manufacturers to foresee problems in advance, therefore minimizing unexpected periods of inactivity and the resulting energy wastage. This proactive approach not only enhances energy efficiency but also contributes to significant cost savings (Qi & Tao, 2018). Moreover, DTs play an important role in supporting sustainable practices within manufacturing. By simulating various energy usage scenarios, manufacturers can identify the most energy-efficient processes and implement them, leading to reduced greenhouse gas emissions. This aligns with global sustainability goals and regulatory requirements aimed at mitigating climate change impacts (Grieves & Vickers, 2016).

The implementation of DT technology in manufacturing is mostly dependent on the use of virtual simulation models to provide exact and complete digital representations of physical systems. These models are critical for improving product processing and assembly, resulting in accurate manufacturing control. This technique consists of numerous important components, including production process simulation, digital production lines, and equipment condition monitoring (He & Bai, 2021). Shao and Helu demonstrated three possible DT applications at different production levels, as indicated in Figure 4.2 (Hananto et al., 2024).

a. **Production Process Simulation**

Production process simulation involves creating a digital representation of the whole manufacturing process. The simulation allows producers to:

- **Visualize and Optimize:** Gain an understanding of the complete manufacturing workflow, identify bottlenecks, and optimize procedures to maximize efficiency.
- **Predict Outcomes:** Forecast the outcomes of various production scenarios to aid decision-making and reduce risk.

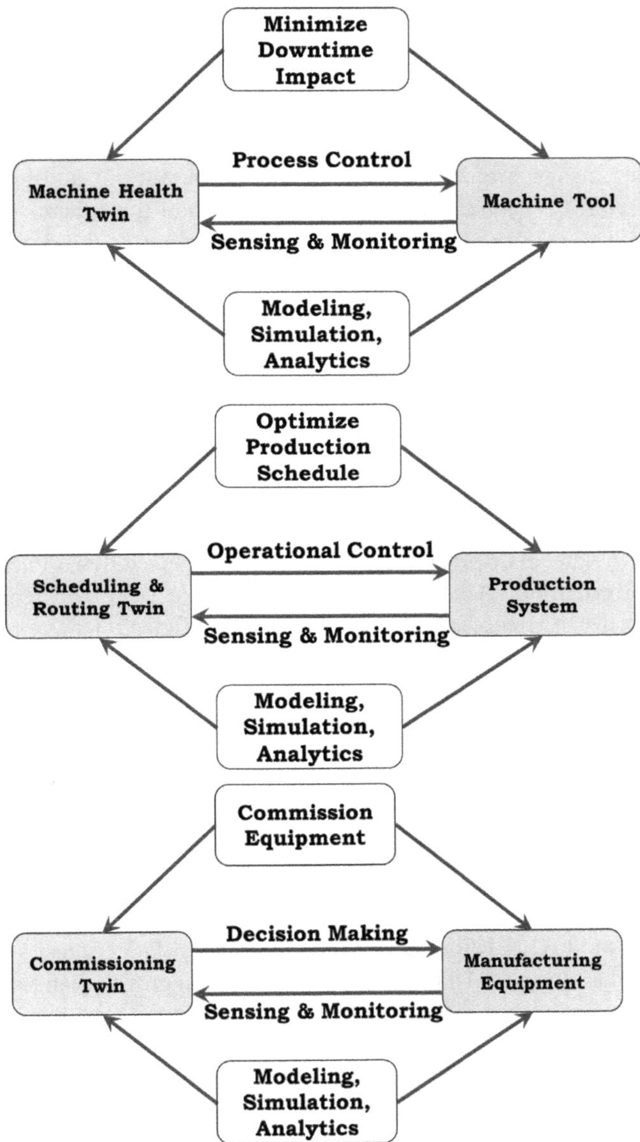

FIGURE 4.2 Application of DTs in industrial processes (Hananto et al., 2024).

- **Validate Designs:** Virtually test and validate new product designs and production procedures before they are implemented on the shop floor.

  Before beginning product production, the complete manufacturing process may be simulated utilizing virtual production techniques. Several novel strategies have been developed to improve this simulation with the integration of resource supply and demand matching, intelligent manufacturing systems, etc.

b. **Digital Production Line**

A digital production line represents the virtual counterpart of the physical production line, providing real-time insights and control over the manufacturing process. Key features include:

- **Synchronization:** Ensures that the virtual and physical production lines are synchronized, allowing real-time monitoring and adjustments.
- **Flexibility:** Facilitates rapid reconfiguration of manufacturing lines to adapt to variations in the volume of production or the designs of the products.
- **Efficiency:** Enhances production efficiency by optimizing resource utilization, reducing downtime, and minimizing waste.

c. **Equipment Status Monitoring**

Monitoring the status of manufacturing equipment is crucial for maintaining optimal production conditions. DTs enable:

- **Real-Time Monitoring:** Continuous tracking of equipment performance and condition using sensor data.
- **Predictive Maintenance:** Predicting equipment failures and scheduling maintenance proactively to prevent unexpected downtimes.
- **Optimization:** Analyzing equipment data to optimize performance, improve efficiency, and extend the lifespan of machinery.

The DT of individual deployed assets provides useful data, but the DT of the manufacturing process is very powerful and captivating. As illustrated in Figure 4.3, this model represents a manufacturing process in the physical world and its digital counterpart. The DT functions as a virtual representation of real-time factory floor activities (Parrott & Lane, 2017). A multitude of sensors distributed throughout the production process gather data across several dimensions, including: Behavioral Characteristics: Attributes of machinery and works in progress, such as thickness, color qualities, hardness, torque, and speeds. Environmental Conditions: Factors like temperature, humidity, and air quality within the factory. These sensors continuously communicate data to the DT application, where it is aggregated and analyzed. DT applications continuously process incoming data streams. Over time, this analysis can reveal unacceptable trends in manufacturing performance compared to an ideal range of tolerable performance. Such insights can prompt investigations and potential adjustments in the physical manufacturing process, demonstrating the interactivity between the physical and digital worlds. This journey highlights the potential of the DT: Continuous Measurement: Thousands of sensors take continuous, non-trivial measurements. Data Streaming: Data is streamed to a digital platform. Real-Time Analysis: Near-real-time analysis is performed to optimize the business process transparently. Figure 4.4 specifically highlights five enabling components: Sensors and Actuators: Collect data from the physical world and execute changes based on DT insights. Integration: Ensures seamless data flow between physical and digital realms. Data: Captured from sensors, it forms the basis for analysis. Analytics: Processes data to derive insights and identify trends. DT Application: Continuously updated with real-time data, enabling informed decision-making and optimization. These components collectively enable the DT to provide powerful and actionable

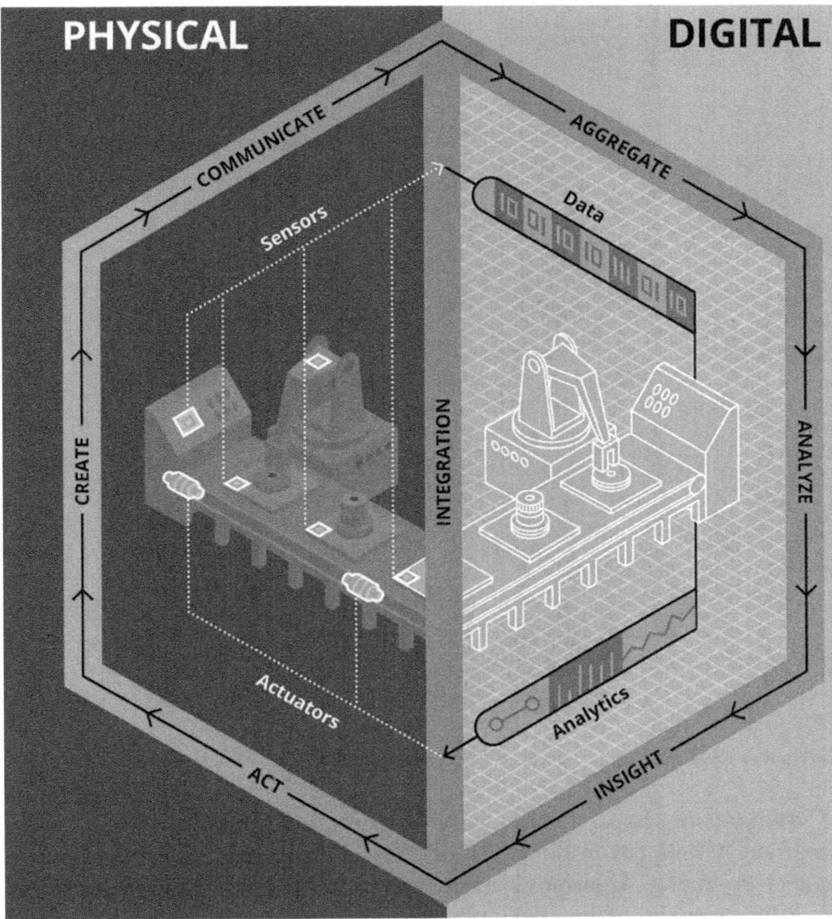

FIGURE 4.3 DT model for a typical manufacturing process (Guo et al., 2023).

insights for enhancing manufacturing processes. By using up-to-the-minute information and sophisticated data analysis techniques, DT helps manufacturers achieve higher efficiency, better quality control, and more adaptive production systems (Evangeline & Anandhakumar, 2020; Guo et al., 2023; Parrott & Lane, 2017).

4.3.1 Digital Twin (DT) Conceptual Architecture

DT conceptual architecture consists of six steps that enable comprehensive modeling of the manufacturing process (Evangeline & Anandhakumar, 2020; Parrott & Lane, 2017). These steps are applicable to various DT configurations. (1) **Create:** Outfit the physical process with sensors measuring operational and environmental data. Transform these measurements into secure digital messages and transmit them to the DT. Integrate additional data from systems like manufacturing execution system (MES), ERP, Computer-Aided Design (CAD) models, and supply chains.

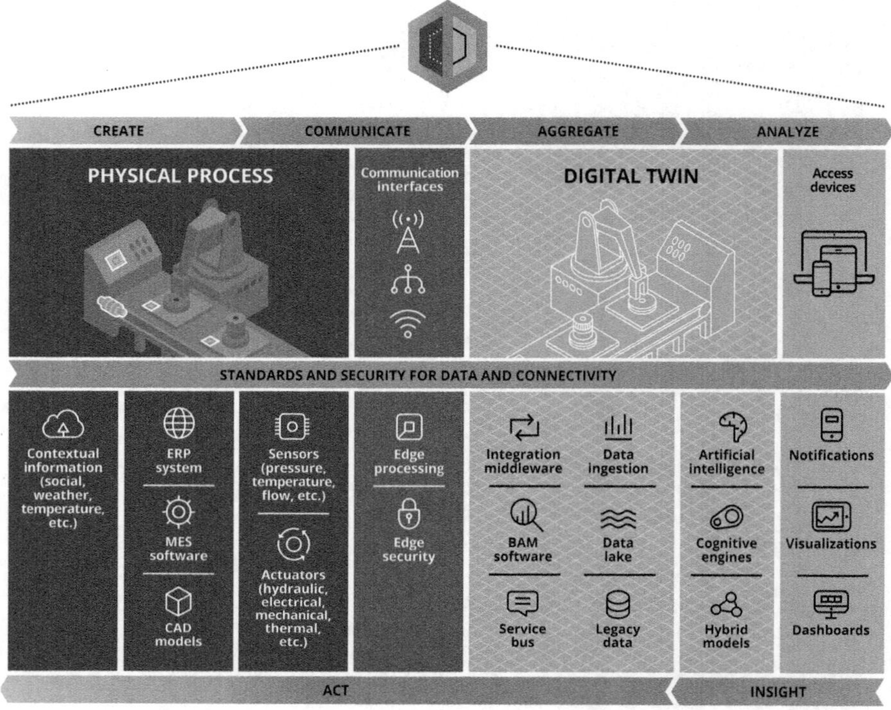

FIGURE 4.4 DT conceptual architecture (Murgod et al., 2023).

(2) **Communicate:** Ensure real time, bidirectional integration between the physical process and the digital platform through **Edge Processing:** Connect and process data near the source. **Communication Interfaces:** Transfer information from sensors. **Edge Security:** Use firewalls, encryption, and certificates to secure data. (3) **Aggregate:** Ingest data into a repository, process it on-premises or in the cloud, and prepare it for analytics using scalable architectures. (4) **Analyze:** Analyze and visualize data using advanced analytics platforms. Develop models to generate insights and guide decision-making. (5) **Insight:** Present insights through dashboards, highlighting performance discrepancies between the DT and the physical process. (6) **Act:** Feed actionable insights back to the physical process through actuators or backend systems, completing the closed-loop connection between the physical and digital worlds.

4.4 DIGITAL TWIN (DT)-DRIVEN SMART MANUFACTURING

Manufacturing is getting smarter, with cognitive intelligence enabling learning, configuration, and execution from physical devices to factory management and production networks. Smart manufacturing means using manufacturing data much more effectively across the whole production and supply chain business. A lot of complex sensor-based data analytics, modeling, and simulation are used in this method to help understand, think about, plan, and manage all parts of the production process in real time. DTs, or virtual versions of actual objects, may be created in smart manufacturing by connecting traditional cyber gateways to the Industrial Internet. By collecting and analyzing data in real time from interconnected objects, Cyberspace DTs may model, simulate, and predict how those objects will behave in dynamic environments. Big Data Processing and AI are two examples of smart technologies that make process optimization and continual extraction of industrial intelligence possible. Internal company operations may benefit from DT technology's enhanced real-time monitoring, predictive maintenance, and process optimization capabilities. More adaptable and personalized production models that can quickly respond to shifts in the market and changes in consumer expectations, as well as improved communication and coordination across the supply chain, all contribute to a rise in inter-business collaboration (Lu et al., 2020).

Emerging smart manufacturing themes include smart production, smart networks, and mass personalization. Cognitively intelligent manufacturing technologies govern production activities in smart production, improving flexibility and efficiency. These self-organizing, networked systems solve new production problems and let people and robots collaborate in immersive industrial settings. Global smart production networks of linked cyber-physical production systems (CPPSs) can adapt to dynamic changes in local production systems and supply chains in near real time. This network of self-optimizing and adaptive systems allows autonomous production configuration and planning, establishing scalable manufacturing positions with economic, environmental, and social advantages. Pull-type mass personalization is replacing push-type mass production in manufacturing. Batch-size-of-one production is efficient and adaptable in advanced facilities that meet factory floor, supply chain, and customer needs. Pervasive manufacturing intelligence lets remote factories and production systems independently perceive, configure, and communicate depending on near-real-time production status and demands. This adaptability allows the quick creation of consumer-specific goods (Lu et al., 2020).

CPS are transformational technologies that govern networked physical resources and computational capabilities. Connection, Conversion, Cyber, Cognition, and Configuration comprise the 5C architecture, which guides industrial CPS development and implementation. MTConnect protocols let sensors or business systems such as Enterprise Resource Planning (ERP), Manufacturing Execution Systems (MES), Supply Chain Management (SCM), and Coordinate Measuring Machines (CMM) to accurately capture machine data during the first step, Connection. In the Conversion stage, prognostics and health management systems assess health values and estimate remaining usable life, giving machines self-awareness. The Cyber stage is a hub where data from linked devices create a network. Analytics extract insights into equipment status and performance, allowing self-comparison and future behavior prediction. Cognition produces system information to facilitate decision-making, including infographic-based maintenance job prioritization. Finally, the Configuration step converts cyberspace input into physical-space actions, enabling machines to self-configure and employ a control system that is adaptive and robust in order to implement corrective measures that are cognition-level and preventive measures (Lee et al., 2015).

MAYA Project (MultidisciplinArY integrated simulAtion and forecasting tools) Project (Negri et al., 2017): For CPS-based manufacturing, the European H2020 project MAYA has improved DT research. This project uses interdisciplinary integrated modeling and forecasting tools with digital continuity and real-world synchronization to optimize manufacturing processes and reduce time to production. It aims to create a plant DT that supports all manufacturing lifecycle stages, from design to dismissal. A key feature of MAYA is the seamless integration of virtual and physical dimensions into a simulation. Semantic meta-data models that comprehensively define CPS properties enable this integration. MAYA relies on the centralized support infrastructure, which offers the following services: The Semantic Meta-Data Model structures data to provide digital continuity across the production system's lifespan. Simulation Framework: Creates a multidisciplinary physical system simulation using many simulation methods and tools. Communication Layer: Connects physical CPS to the digital world for real-time synchronization and updates with massive field data. MAYA found the key DT features for Industry 4.0 production systems. DTs are virtual production systems that can run many simulation disciplines. It synchronizes virtual and real systems using sensed data, linked smart devices, mathematical models, and real-time data elaboration. Industry 4.0 manufacturing systems rely on the DT to predict and optimize production system behavior at each lifecycle step in real time. Industry 4.0 technologies allow this capacity to align with academic viewpoints.

4.4.1 Use of Digital Twin (DT) in Industry 4.0—Manufacturing Sector

The use of DT technology in Industry 4.0 has revolutionized traditional procedures, resulting in significant improvements in effectiveness, output, and creativity. The influence of DT in the manufacturing sector includes cutting tools, three-dimensional (3D) printing, additive manufacturing (AM), dynamic scheduling, and data management.

a. **Cutting Tools:** The use of DT technology has significantly transformed the process of managing and enhancing the performance of cutting tools. Manufacturers may acquire valuable insights on tool performance and wear by creating virtual versions of these tools. This allows for improved process planning and ongoing enhancements. In a study, the integration of real-time data with virtual models in order to enhance machining solutions and the ways in which DT improves the management of cutting tools using ISO 13399 standards were demonstrated (Botkina et al., 2018). Communication and control of cutting tool information are crucial for industrial efficiency and accuracy. DT technology helps handle cutting tool data, including: Each tool's version, definition, attributes, and relationships are included. A virtual representation called a DT incorporates this data, assuring accuracy and consistency across platforms. Tool assembly information: Definitions, attributes, and instructions. DTs provide extensive data for tool assembly and use when integrated with CAD, Computer-Aided Manufacturing (CAM), and Computer Numerical Control (CNC) software. External Document References: DTs connect to 3D models of tools or tool assemblies for fast access to design and assembly manuals. Multifunctionality: The DT environment truly represents and makes multifunctional tools available

(Botkina et al., 2018). Xie et al. (2021) provide a system that utilizes DT technology to manage data flow in cutting tools. The approach specifically tackles issues connected to data analysis and service integration.

b. **3D Printing and AM:** By incorporating DTs into the processes of 3D printing and AM, the issues of material restrictions and the occurrence of defects are effectively tackled. DTs enable the implementation of manufacturing processes that are both more precise and cost-efficient. A detailed DT framework for 3D printing with the goal of reducing flaws and expediting the process of verifying the quality of parts was proposed (Mukherjee & DebRoy, 2019). By incorporating DT technology into 3D printing, production reliability, safety, and precision are greatly improved. This is achieved via the optimization of manufacturing processes and the guarantee of high-quality goods. A 3D printing DT consists of many essential elements: the mechanical model, sensor and control model, statistical model, and big data. The mechanistic model utilizes well-established engineering and metallurgical concepts to predict important characteristics such as transient temperature fields, solidification morphology, grain structure, and defect susceptibility. The software simulates the interactions between heat sources, such as lasers and electron beams, and metallic materials. It utilizes several techniques, including analytical approaches, Finite Element Method heat conduction models, heat transfer and fluid flow models, and powder scale models. Although these models may need significant computer resources, they are essential for accurately forecasting temperature distributions, cooling speeds, and deposit shapes. The sensing and control model utilizes sensors such as infrared cameras and acoustic emission systems to continuously monitor and regulate the 3D printing process. This real-time monitoring ensures that process variables are maintained within ideal ranges, hence preventing any faults. The statistical model utilizes prior experiments and extensive data to enhance predictions, specifically targeting the constraints of mechanistic models. Big data includes sensor data, test results, mechanistic models, and relevant literature. These elements together increase the accuracy of models and aid in continual enhancements of the 3D printing process. The combination of models and data enables improved quality control, process optimization, and informed decision-making, leading to progress in the manufacturing industry. These contributions demonstrate the ability of DTs to address constraints in AM by providing accurate and efficient production techniques (Mukherjee & DebRoy, 2019).

c. **Dynamic Scheduling:** The process of scheduling tasks or activities in a flexible and adaptable manner, taking into account real-time conditions and priorities. Utilizing real-time data, DTs are crucial in optimizing job-shop scheduling and improving operational efficiency via simulation and analysis. Work delays, urgent work arrivals, and processing time variations might influence scheduling in addition to machine availability. The DT-based disturbance detection system compares the PE to the virtual entity (VE), a dynamic reference, to identify disturbances and anticipate negative consequences early to facilitate prompt rescheduling. DT-based performance assessment mimics machine indicators including utility rate, stress level, and energy consumption utilizing multi-dimensional virtual models in the VE and real-time PE data before rescheduling. Before each assessment, the VE receives real-time PE data (e.g., processing time, spindle speed, and cutting force) to update model parameters (Zhang et al., 2021).

d. **Maintenance:** The use of DTs for maintenance is becoming more prevalent, providing substantial advantages in operational efficiency and safety. Errandonea et al. (2020) conducted a thorough analysis of the use of DTs in maintenance, focusing on ideas, tactics, and potential areas for further study. Their work showcased the capacity of DTs to influence firm operations, including the administration of manufacturing lines and the assurance of worker safety. These advancements highlight the significant influence of DTs on maintenance, providing improved predictive capacities and operational insights.

4.4.2 Real-Time Tool Condition Monitoring (TCM)

Cutting tools are important parts of making because they remove material from workpieces, shape them, and cut them. Cutting tools, on the other hand, often break when they are used in places with too much force, stress, and high temperatures. This can damage workpieces, stop production, and cost the company money. Tool condition monitoring (TCM) is an important way to cut down on machine tool downtime and make production more reliable. TCM is direct or indirect. Downtime is required to monitor tool wear using a tool makers microscope or optical microscope in the direct approach. Instead, the indirect approach estimates tool wear using empirical correlations between tool wear and measurements, making it better for real-time TCM but less exact. Most indirect TCM techniques analyze data to determine acoustic emission, cutting force, vibration, sound, and power. Signal feature-based techniques are useful, but feature extraction is difficult and inflexible. Model feature-based TCM solves these problems. These techniques create dynamic process models from signals and extract frequency response characteristics as TCM features, making them more flexible and simpler than signal feature-based methods. Real-time TCM may be possible with DTs' real-time data-driven modeling, frequency domain analysis, and diagnostics (Liu et al., 2024). This system monitors complicated cutting tool conditions. The suggested framework has three parts:

- **Physical Product**: The cutting tool, workpiece, milling fixture, spindle dynamics, and workpiece-cutting tool interactions
- **Virtual Product**: Shows the dynamic interaction between vibration data and model frequency features (MFFs)-based diagnostics for cutting tool abnormalities in real time.
- **Connections to Data Flow**: Real-time vibration data and machine tool numerical controller (NC) inputs give machine tool dynamics and machining process information.

4.4.2.1 Implementation

The implementation involves integrating sensors with machine tools, setting up data transmission systems, and developing dynamic models to analyze vibration data. Key factors include ensuring appropriate sampling rates and minimizing latency.

4.4.2.2 Framework

- **Live Data:** Uses machine tool sensors' real-time operating data with precise sample rates and delay.
- **Digital Coupling:** Includes data collection, transmission, and reception for real-time physical state representation.
- **State:** NC code-provided cutting tool and machining feature execution status and locations.
- **Physical Product:** Sensors measure spindle, cutting tool, workpiece, and milling fixture vibration and dynamics.
- **Virtual Product:** A TCM diagnosis method, frequency domain characteristics, and live updated data-driven modeling of the physical product's behavior in real time are included.
- **Functional Output:** Determines cutting tool health and replacement or usefulness.

4.4.3 Production Scheduling

Manufacturing relies on production scheduling to allocate resources and sequence operations to minimize make span, costs, and work tardiness. With Industry 4.0, static scheduling systems are being replaced with dynamic ones that are more flexible and adaptable to disturbances.

Dynamic scheduling allows for the flexible adjustment of production plans, either partially or completely, in order to minimize the negative effects of interruptions (Ouahabi et al., 2024). The main categories of dynamic scheduling include reactive scheduling, robust scheduling, and predictive-reactive scheduling.

- **Reactive Scheduling:** It minimizes production disruptions by adapting to real-time changes. A reactive scheduling technique bases scheduling on production rather than a schedule. This technique may react fast to disruptions but lacks stability and global ideal answers due to its reliance on real-time information.
- **Reliable Scheduling:** It anticipates disturbances to produce a resilient starting plan. This method is more steady and disturbance-tolerant. However, unforeseen occurrences may invalidate the timeframe.
- **Proactive-Reactive Scheduling:** Reactive and robust scheduling are combined in predictive-reactive scheduling. It can predict disruptions, adjust plans, and respond in real time. This hybrid approach blends stability and agility to manage unforeseen changes.

Zero-defect manufacturing (ZDM) refers to a manufacturing process that aims to provide flawless products by eliminating any defects or errors. ZDM signifies a significant paradigm change in quality management, prioritizing the prevention of errors rather than only relying on their detection and subsequent treatment. The DT-ZDM scheduling method includes fault detection, prediction, and preventive tactics inside dynamic scheduling procedures. Manufacturing is focused on the needs and preferences of humans. Industry 5.0 prioritizes a human-centered approach, where technology is designed to accommodate the demands and capacities of people. Integrating DT into dynamic production scheduling has great potential for transforming smart manufacturing. DT efficiently addresses major challenges in modern manufacturing environments by enhancing sustainability, quality, and the incorporation of human factors. These advancements improve effectiveness and also contribute to manufacturing systems that are flexible and resilient, capable of responding to dynamic changes and improving overall performance (Ouahabi et al., 2024).

4.4.4 Additive Manufacturing (AM)

AM uses DT technology to solve process repeatability, material waste, and component characterization issues. DT simulates the AM process for real-time monitoring and optimization. DT improves process repeatability and waste reduction via predictive maintenance and process optimization using real-time data and virtual models. AM processes are improved by big data analytics, AI, and IoT. DT applications benefit from physics-based, data-driven, and hybrid modeling, boosting quality and lowering costs. Despite advances, standardization and universal reference models remain issues (Chen et al., 2024).

The role of DT in AM has been comprehensively reviewed and summarized by Jin et al. (2024) and is listed below.

4.4.4.1 Design/Prototyping Improvement

Engineers may model and test designs before production using DTs, speeding up design cycles and prototyping. This feature speeds development and identifies difficulties early.

- **Virtual Prototyping:** Tests AM component performance, durability, and manufacturability using virtual models.
- **Design Optimization:** Improves component functioning and reduces material waste by adjusting designs based on simulation findings.

4.4.4.2 Process Monitoring and Control

Real-time monitoring and control are essential in AM to guarantee the quality and consistency of the printed parts. Digital duplicates facilitate the immediate adjustment of the manufacturing process and enable continuous observation.

- **Real-Time Data Integration:** Gathers and consolidates data from sensors to monitor process parameters, including temperature, pressure, and material flow.
- **Adaptive Control:** Dynamically modifies process parameters in response to real-time data to ensure optimal conditions and improve the quality of the part.

4.4.4.3 Quality Assurance and Predictive Maintenance

DTs contribute significantly to quality assurance and maintenance by providing insights into the performance and condition of manufacturing equipment.

- **Defect Detection:** Uses real-time data and simulations to detect and predict potential defects in parts before they occur.
- **Predictive Maintenance:** Analyzes equipment performance data to forecast maintenance needs, reducing downtime and extending equipment life.

4.4.4.4 Workflow Optimization

DTs streamline the entire AM workflow by integrating design, production, and post-production processes into a cohesive system.

- **Process Simulation:** Allows for the simulation of the entire AM workflow to identify bottlenecks and optimize resource allocation.
- **Supply Chain Integration:** Connects with supply chain management systems to optimize material flow, inventory, and production scheduling.

Future developments in DT technology for AM will improve surrogate modeling, polish sensor technologies, and provide consistent ontologies. While worldwide partnerships will concentrate on multiscale modeling and real-time applications to hasten adoption and invention, integration with lifecycle data will boost predictive maintenance and problem detection (Jin et al., 2024).

4.5 ENABLING TECHNOLOGIES FOR DIGITAL TWIN (DT)

Several supporting technologies help create, operate, and optimize DTs for industrial processes. This section provides an overview of these enabling technologies, their functions, and their effects on DT manufacturing. It also summarizes the key enabling technologies for cognizing and controlling the physical world, managing DT data, providing DT services, and ensuring robust connections that support DT development, implementation, and optimization (Qi et al., 2021).

4.5.1 Physical World Recognition and Control Technologies

Complex physical events must be measured and modeled virtually to recognize the physical world. This method uses sophisticated measuring instruments. Precision engineering and manufacturing need laser measuring instruments to accurately capture measurements and surface details. Image recognition technologies analyze visual data using cameras and advanced algorithms to enable thorough measurement and inspection. Conversion measurement techniques that convert physical features into digital signals enable physical data integration with virtual models. Aerospace and microelectronics need exact standards; hence, micro- and nano-level measuring methods are essential (Gao et al., 2015).

Real-time data collection is essential for realistic virtual representations of physical processes. Sensors continually monitor torque, pressure, displacement speed, and ambient factors. Evolution models, until measurement technologies for precision positioning are used in industry, forecast performance by observing PE changes over time. Structural analysis models assess structural integrity and performance, whereas data-driven fault prediction approaches identify failures and maintenance requirements. Power and drive systems must be managed to control physical things. Hydraulic, electric, and fuel systems power actuators using diverse mechanical movements. Servo and shaftless drives provide accurate mechanical motions. Effective process control uses planning, design, and optimization technologies. Operation precision is ensured by electrical, programmable, and network controls. Big data analytics and machine vision improve perception and control. Big data analytics can handle vast quantities, different kinds, and high velocities of data, providing useful insights. Machine vision uses neurobiology, image processing, and pattern recognition to detect and quantify. Smart robots and innovative construction materials enhance DTs (Adamson et al., 2017; Angrish et al., 2017).

4.5.2 DT-Data Management Tools

Collection, transmission, storage, processing, and visualization are crucial to DT data management. Data gathering uses sensors, IoT devices, Application Programming Interface (APIs), and databases. Wireless and cable network technologies are needed for efficient data flow. NoSQL, NewSQL, and cloud storage can scale and adapt to huge datasets. Data processing includes preparation, analysis, and deep learning. Data analysis requires preprocessing. Statisticians and neural networks analyze data in detail, while deep learning improves analytics and pattern recognition. Raw-data, feature-level, and decision-level data fusion give complete insights from many sources (Wu et al., 2017). AI and random methods aid these procedures. Data visualization, including histograms, charts, and dashboards, clarifies results. Parallel methods and dimensionality reduction may enhance data understanding in visualization (Qi et al., 2021).

4.5.3 DT Service Enabling Technologies

For development, maintenance, and on-demand consumption, DT services use many technologies. Service generation requires virtualization and encapsulation technologies like service oriented architecture (SOA) and web services, as well as sensors and middleware for resource monitoring and assessment. Service management technologies seek, match, and evaluate service quality to assure performance and dependability. On-demand service consumption requires transaction management technology for automatic matching and monitoring (Qi et al., 2021). Knowledge services use decision trees and neural networks to acquire and store knowledge in DT operations. Reuse this knowledge with service platforms. Service publication, querying, and security management are supported by industrial IoT systems. Device, network, and information security are needed to secure DT systems.

4.5.4 DT-Connection Technologies

DT systems need connections for real-time component interaction. Seamless integration requires six connections such as (i) connections between, physical entities and virtual models, (ii) connection between physical entities and data, (iii) connection between physical entities and services, (iv) connection between virtual models and data, (v) connection and between virtual models and services, (vi) connection between virtual models and services. Physical items must be accurately represented via Radio-Frequency Identification (RFID) and IoT. Unified communication interfaces and protocols let virtual models and data communicate. Virtual Reality (VR), Augmented Reality (AR), and Mixed Reality (MR) improve DT user engagement. Device, network, and information security are needed to protect DT systems (Qi et al., 2021). New connection theories and standards will resolve interface and protocol differences. As data traffic rises, green communication methods will be needed to manage energy and bandwidth.

4.6 CASE STUDY: FESTO CYBER-PHYSICAL SMART FACTORY

In this case study, the Festo Cyber-Physical Smart Factory reported by Onaji et al. (2022) is discussed. Festo's Cyber-Physical Smart Factory is a cutting-edge illustration of Industry 4.0 integration, specifically engineered to foster the acquisition and cultivation of expertise in intelligent industrial automation. This compact CPPS incorporates a variety of resources into a unified system, utilizing distributed control, RFID technology, and sensors to improve industrial automation. A comprehensive platform for process monitoring and experimental analysis is provided by the MES that manages the factory (Jirsa, 2020).

4.6.1 System Overview

The Festo Cyber Physical (CP) smart factory has a network system that uses the Transmission Control Protocol/Internet Protocol (TCP/IP), Ethernet, and open platform communications unified architecture (OPC_UA) protocols. This setup makes it easy for system parts to share information with each other. The plant has two production islands, a supply network with transports and elevators, and a robotino that can drive itself. The system uses a method called "distributed control," in which a programmable logic controller (PLC) runs each part (Table 4.1).

TABLE 4.1
Festo CP Smart Manufacturing Stations and Their Respective Functions

Station	Task
Top case (Station 1)	Place the top cover of the phone on the carrier
Measuring (Station 4)	Inspect the workpiece on the carrier
Bottom case (Station 5)	Load the carrier with the bottom cover of the phone
Press (Station 6)	Couple the top and bottom covers by pressing them together
Heat tunnel (Station 7)	Heat the workpiece to a predefined temperature
Output station (Station 8)	Remove the finished product from the production line
Bridges (Station 2 and 3)	Transfer workpieces between workstations
Robotino	Transfer products between workstations

4.6.2 Approach

A structured approach was employed in the development of the digital counterpart for the Festo CP smart factory.

a. **Design of Conceptual Models:** Specified the scope of the system and the objectives of the endeavor. Developed a conceptual model that delineates the system architecture, product attributes, and processes for digital assessment.
b. **Development of Virtual Models:** Utilized Siemens NX for 3D CAD modeling and Tecnomatix for discrete event simulations (DES) to develop simulation models for processes and products. Developed OPC_UA interfaces to facilitate bidirectional communication between virtual and physical systems.
c. **Development of a Control Panel:** Added control elements, including push buttons that are linked to virtual model control instructions and data display elements.
d. **Development of the Data Layer:** Integrated data storage components, such as tables and variables, to facilitate the visualization and analysis of operational and virtual data.
e. **The Development of Intelligent Layers:** Developed algorithms to facilitate the generation and processing of virtual data, thereby facilitating the administration of systems and the acquisition of real-time insights.
f. **Validation and Verification:** Conducted continuous testing and debugging to guarantee that the project's designs were accurate and in accordance.

4.6.3 DT Model for Discrete Event Simulation (DES)

The Festo CP smart factory's DES digital counterpart was developed with predetermined behaviors and interactions, which were implemented through control algorithms to regulate material flow. The dynamic operations of the physical system are reflected in the digital counterpart architecture and process flow. The digital counterpart of the product simulates the upper and lower cases of a phone. RFID technology, which is affixed to carriers, monitors and updates product data in real time as it progresses through the production line. The product DT's geometric and information models are illustrated in Figure, which guarantees optimal synchronization between the physical product and its digital representation (Figure 4.5).

4.6.4 Benefits Observed

The DT enhances the system's research and experimentation capabilities, enabling diagnostic/predictive studies, business case analysis, evaluation of changes, optimum production configurations, and monitoring. It also offers some flexibility in teaching/training. The DES DT, accessible offline/online, addresses the space and accessibility constraints faced by many students in the physical system. Industry 4.0 ideas and technology may be taught securely, particularly in scenarios like COVID-19 where

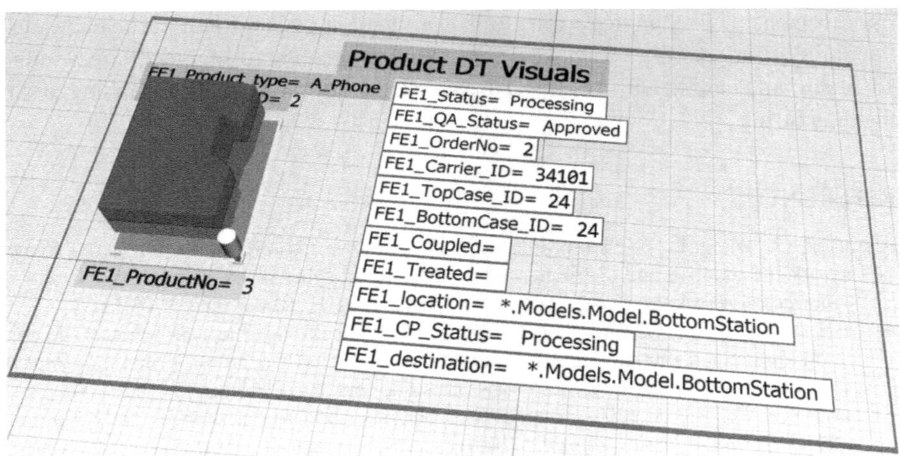

FIGURE 4.5 Product DT visuals in Festo CP smart factory (Onaji et al., 2022).

physical distance is crucial for safety. The digital duplicate will be upgraded to an interactive and immersive model with sophisticated predictive capabilities in the subsequent phase.

4.6.5 Future Work

The Festo CP smart factory project's subsequent phase is designed to enhance the current DT into a predictive model that is more interactive and immersive. The complexities and vagaries that are encountered in real-world manufacturing environments will be more faithfully represented by this upgraded DT. Advanced decision-making capabilities will be integrated into the improved model to improve the simulation and prediction of operational scenarios.

4.7 CONCLUSION

DT technology is revolutionizing production with unparalleled monitoring, prediction, and optimization. It uses CPS, IoT, and AI to connect the physical and digital worlds, boosting innovation and efficiency in Industry 4.0. It lets manufacturers mimic, monitor, and improve product lifecycle processes by representing physical items virtually. Proactive management with real-time analytics and predictive maintenance reduces downtime and energy losses. DT facilitates sustainable production by modeling scenarios to reduce energy consumption and environmental effects. DT provides dynamic production scheduling flexibility and responsiveness. They alter production plans in real time, utilizing reactive, robust, and predictive-reactive scheduling to resolve disturbances and optimize performance. It also helps ZDM improve quality and efficiency by integrating problem detection, prediction, and prevention into scheduling. It assists in AM, 3D printing, and TCM. DT optimizes processes and solves material restrictions to provide high-quality results and continual improvement. DT technology transforms production by improving efficiency, sustainability, and quality. It makes conventional manufacturing techniques more adaptable and intelligent, meeting current production needs and advancing smart manufacturing.

REFERENCES

Adamson, G., Wang, L., & Moore, P. (2017). Feature-based control and information framework for adaptive and distributed manufacturing in cyber physical systems. *Journal of Manufacturing Systems*, *43*, 305–315. https://doi.org/10.1016/j.jmsy.2016.12.003

Angrish, A., Starly, B., Lee, Y. S., & Cohen, P. H. (2017). A flexible data schema and system architecture for the virtualization of manufacturing machines (VMM). *Journal of Manufacturing Systems*, *45*, 236–247. https://doi.org/10.1016/j.jmsy.2017.10.003

Attaran, M., & Celik, B. G. (2023). Digital twin: benefits, use cases, challenges, and opportunities. *Decision Analytics Journal*, *6*, 100165. https://doi.org/10.1016/j.dajour.2023.100165

Botkina, D., Hedlind, M., Olsson, B., Henser, J., & Lundholm, T. (2018). Digital twin of a cutting tool. *Procedia CIRP*, *72*, 215–218. https://doi.org/10.1016/j.procir.2018.03.178

Chen, Z., Surendraarcharyagie, K., Granland, K., Chen, C., Xu, X., Xiong, Y., Davies, C., & Tang, Y. (2024). Service oriented digital twin for additive manufacturing process. *Journal of Manufacturing Systems*, *74*, 762–776. https://doi.org/10.1016/j.jmsy.2024.04.015

Errandonea, I., Beltrán, S., & Arrizabalaga, S. (2020). Digital twin for maintenance: a literature review. *Computers in Industry*, *123*, 103316. https://doi.org/10.1016/j.compind.2020.103316

Evangeline, P., & Anandhakumar, P. (2020). Digital twin technology for "smart manufacturing." *Advances in Computers*, *117*, 35–49. https://doi.org/10.1016/bs.adcom.2019.10.009

Fuller, A., Fan, Z., Day, C., & Barlow, C. (2020). Digital twin: enabling technologies, challenges and open research. *IEEE Access*, *8*, 108952–108971. https://doi.org/10.1109/ACCESS.2020.2998358

Gao, W., Kim, S. W., Bosse, H., Haitjema, H., Chen, Y. L., Lu, X. D., Knapp, W., Weckenmann, A., Estler, W. T., & Kunzmann, H. (2015). Measurement technologies for precision positioning. *CIRP Annals - Manufacturing Technology*, *64*(2), 773–796. https://doi.org/10.1016/j.cirp.2015.05.009

Glaessgen, Edward H., and D. S. Stargel. 2012. "The Digital Twin Paradigm for Future NASA and U.S. Air Force Vehicles." *53rd AIAA/ASME/ASCE/AHS/ASC Structures, Structural Dynamics and materials Conference*, NF1676L-13293, Honolulu, HI.

Grieves, M. (2014). Digital Twin: Manufacturing Excellence through Virtual Factory Replication. *White Paper*, March, 1–7. https://www.researchgate.net/publication/275211047_Digital_Twin_Manufacturing_Excellence_through_Virtual_Factory_Replication

Grieves, M., & Vickers, J. (2016). Digital twin: mitigating unpredictable, undesirable emergent behavior in complex systems. In: Kahlen, J., Flumerfelt, S., Alves, A. (eds) *Transdisciplinary Perspectives on Complex Systems: New Findings and Approaches*, 85–113. Springer, Cham. https://doi.org/10.1007/978-3-319-38756-7_4

Grübel, J., Thrash, T., Aguilar, L., Gath-Morad, M., Chatain, J., Sumner, R. W., Hölscher, C., & Schinazi, V. R. (2022). The Hitchhiker's guide to fused twins: a review of access to digital twins in situ in smart cities. *Remote Sensing*, *14*(13), 14133095. https://doi.org/10.3390/rs14133095

Guo, Y., Klink, A., Bartolo, P., & Guo, W. G. (2023). Digital twins for electro-physical, chemical, and photonic processes. *CIRP Annals*, *72*(2), 593–619. https://doi.org/10.1016/j.cirp.2023.05.007

Hananto, A. L., Tirta, A., Herawan, S. G., Idris, M., Soudagar, M. E. M., Djamari, D. W., & Veza, I. (2024). Digital twin and 3D digital twin: concepts, applications, and challenges in industry 4.0 for digital twin. *Computers*, *13*(4), 13040100. https://doi.org/10.3390/computers13040100

He, B., & Bai, K. J. (2021). Digital twin-based sustainable intelligent manufacturing: a review. *Advances in Manufacturing*, *9*(1), 1–21. https://doi.org/10.1007/s40436-020-00302-5

Javaid, M., Haleem, A., & Suman, R. (2023). Digital twin applications toward industry 4.0: a review. *Cognitive Robotics*, *3*, 71–92. https://doi.org/10.1016/j.cogr.2023.04.003

Jin, L., Zhai, X., Wang, K., Zhang, K., Wu, D., Nazir, A., Jiang, J., & Liao, W. H. (2024). Big data, machine learning, and digital twin assisted additive manufacturing: a review. *Materials and Design*, *244*, 113086. https://doi.org/10.1016/j.matdes.2024.113086

Jirsa, I. J. (2020). The introduction of CP Factory production line, ideal technological platform for study and research in the field of automation. *International Scientific Journals*, *234*(5), 233–234.

Kritzinger, W., Karner, M., Traar, G., Henjes, J., & Sihn, W. (2018). Digital twin in manufacturing: a categorical literature. *IFAC-PapersOnLine*, *51*(11), 1016–1022. https://doi.org/10.1016/j.ifacol.2018.08.474

Lee, J., Bagheri, B., & Kao, H. A. (2015). A cyber-physical systems architecture for industry 4.0-based manufacturing systems. *Manufacturing Letters*, *3*, 18–23. https://doi.org/10.1016/j.mfglet.2014.12.001

Lim, K. Y. H., Zheng, P., & Chen, C. H. (2020). A state-of-the-art survey of digital twin: techniques, engineering product lifecycle management and business innovation perspectives. *Journal of Intelligent Manufacturing*, *31*(6), 1313–1337. https://doi.org/10.1007/s10845-019-01512-w

Liu, Z., Lang, Z. Q., Gui, Y., Zhu, Y. P., & Laalej, H. (2024). Digital twin-based anomaly detection for real-time tool condition monitoring in machining. *Journal of Manufacturing Systems*, *75*, 163–173. https://doi.org/10.1016/j.jmsy.2024.06.004

Lu, Y., Liu, C., Wang, K. I. K., Huang, H., & Xu, X. (2020). Digital twin-driven smart manufacturing: connotation, reference model, applications and research issues. *Robotics and Computer-Integrated Manufacturing*, *61*, 101837. https://doi.org/10.1016/j.rcim.2019.101837

Mukherjee, T., & DebRoy, T. (2019). A digital twin for rapid qualification of 3D printed metallic components. *Applied Materials Today*, *14*, 59–65. https://doi.org/10.1016/j.apmt.2018.11.003

Murgod, T. R., Sundaram, S. M., Mahanthesha, U., & Murugesan, P. (2023). A survey of digital twin for industry 4.0: benefits, challenges and opportunities. *SN Computer Science*, *5*(1), 76. https://doi.org/10.1007/s42979-023-02363-2

Negri, E., Fumagalli, L., & Macchi, M. (2017). A review of the roles of digital twin in CPS-based production systems. *Procedia Manufacturing*, *11*, 939–948. https://doi.org/10.1016/j.promfg.2017.07.198

Onaji, I., Tiwari, D., Soulatiantork, P., Song, B., & Tiwari, A. (2022). Digital twin in manufacturing: conceptual framework and case studies. *International Journal of Computer Integrated Manufacturing*, *35*(8), 831–858. https://doi.org/10.1080/0951192X.2022.2027014

Ouahabi, N., Chebak, A., Kamach, O., Laayati, O., & Zegrari, M. (2024). Leveraging digital twin into dynamic production scheduling: a review. *Robotics and Computer-Integrated Manufacturing*, *89*, 102778. https://doi.org/10.1016/j.rcim.2024.102778

Parrott, A., & Lane, W. (2017). *Industry 4.0 and the Digital Twin*. Deloitte University Press, New York, 1–17. https://dupress.deloitte.com/dup-us-en/focus/industry-4-0/digital-twin-technology-smart-factory.html

Qi, Q., & Tao, F. (2018). Digital twin and big data towards smart manufacturing and industry 4.0: 360 degree comparison. *IEEE Access*, *6*, 3585–3593. https://doi.org/10.1109/ACCESS.2018.2793265

Qi, Q., Tao, F., Hu, T., Anwer, N., Liu, A., Wei, Y., Wang, L., & Nee, A. Y. C. (2021). Enabling technologies and tools for digital twin. *Journal of Manufacturing Systems*, *58*, 3–21. https://doi.org/10.1016/j.jmsy.2019.10.001

Tao, F., Zhang, H., Liu, A., & Nee, A. Y. C. (2019). Digital twin in industry: state-of-the-art. *IEEE Transactions on Industrial Informatics*, *15*(4), 2405–2415. https://doi.org/10.1109/TII.2018.2873186

Uhlemann, T. H. J., Lehmann, C., & Steinhilper, R. (2017). The digital twin: realizing the cyber-physical production system for industry 4.0. *Procedia CIRP*, *61*, 335–340. https://doi.org/10.1016/j.procir.2016.11.152

Wu, D., Liu, S., Zhang, L., Terpenny, J., Gao, R. X., Kurfess, T., & Guzzo, J. A. (2017). A fog computing-based framework for process monitoring and prognosis in cyber-manufacturing. *Journal of Manufacturing Systems*, *43*, 25–34. https://doi.org/10.1016/j.jmsy.2017.02.011

Xie, Y., Lian, K., Liu, Q., Zhang, C., & Liu, H. (2021). Digital twin for cutting tool: modeling, application and service strategy. *Journal of Manufacturing Systems*, *58*(PB), 305–312. https://doi.org/10.1016/j.jmsy.2020.08.007

Zhang, M., Tao, F., & Nee, A. Y. C. (2021). Digital twin enhanced dynamic job-shop scheduling. *Journal of Manufacturing Systems*, *58*(PB), 146–156. https://doi.org/10.1016/j.jmsy.2020.04.008

5 Energy Utilisation in Food Processing Industries

Prashant Kumar Jangde, T.V. Arjunan, and Pradeep Patanwar

5.1 OVERVIEW

From farm to plate, the food supply chain is a convoluted process of which food processing plays a vital role and mainly involves conversion of agricultural products into consumable food as well as transforming one food item to another such that it gives more value to the consumers, which is called value-added products [1]. Sometimes, it is mainly used for increasing the life of the product either by storing them at certain atmosphere or by reducing its moisture content. Numerous energy-intensive processes, including heating, cooling, drying, refrigeration, and pasteurisation, are involved in the production of food. The majority of energy is used in the numerous operations, which include baking, pasteurisation, brewing, drying, dehydrating, boiling, chilling, pureeing, cooking, and washing and cleaning. These procedures are necessary to guarantee food safety, maintain food quality, and increase shelf life. A number of procedures are often included in food processing in order to maximise the commercial value of goods. The United Nations Conference on Trade and Development (UNCTAD) reported that between 2000 and 2021, the value of food trade worldwide increased by 350% to $1.7 trillion. Food now makes up almost 8% of all international trade in goods, up from 6% in 2000. Developed economies import more processed food than developing economies do; for example, 48% of all food imports come from developed economies, compared to 35% from the latter. The food enterprises include energy-intensive processes, and energy-related expenses make up 20% to 50% of total production costs [2]. Agriculture, transportation, food processing, and food handling account for the four main stages of energy consumption in food production. The food industry's increased processing levels due to the growing demand for prepared meals have resulted in higher energy use [3] (Figure 5.1).

5.2 TYPES OF FOOD INDUSTRY

Food can be categorised into three groups based on its degree of processing: minimally processed, moderately processed, and highly processed. Minimally processed foods are ones that still have the majority of their natural characteristics such as milk, eggs, and packaged fruits. The next stage of processing, known as moderately processed items, involves preserving, enhancing nutrients, and processing food for freshness. These include items such as frozen vegetables and fruits, tomato sauce,

DOI: 10.1201/9781003452072-5

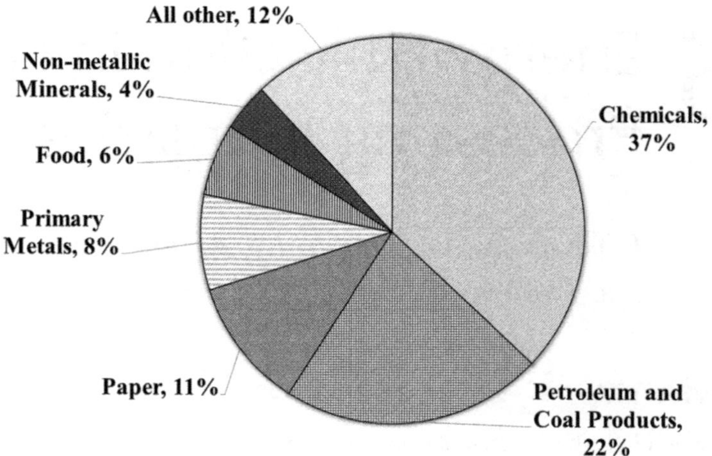

FIGURE 5.1 Percentage of energy used by various industrial sectors, 2018 (EIA).

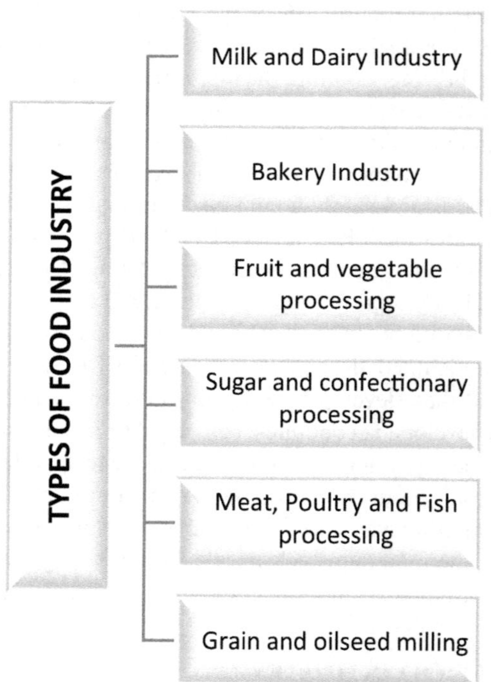

FIGURE 5.2 Types of food industries.

and cheeses. Highly processed food items include ready-to-eat packed food item with minimal dietary fibre content and high levels of added and total sugars such as pizza, alcoholic beverages, and ice cream. [4]. Although there are many other food sectors, the following are some of the primary ones that this study focuses on (Figure 5.2).

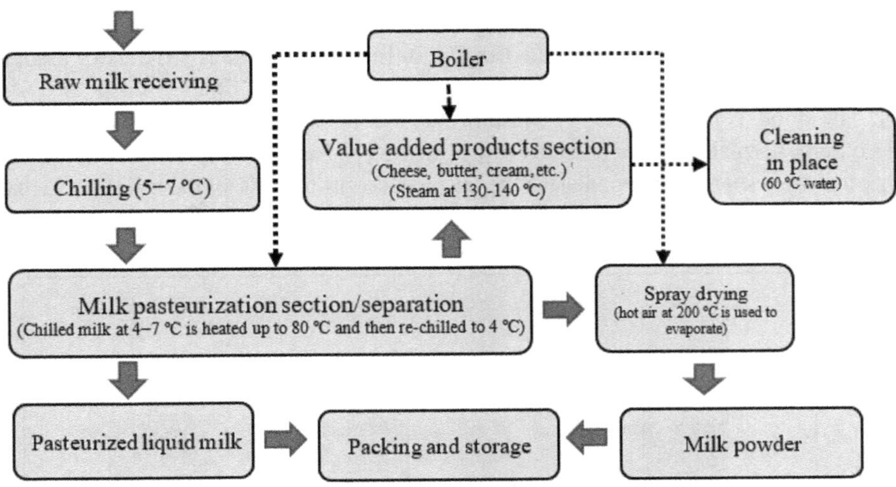

FIGURE 5.3 Flowchart of the milk processing unit [7].

5.2.1 Milk and Dairy Industry

Raw milk is used to make several different products in the dairy processing industry, including yogurt, cheese, butter, and powdered milk and whey. In the food processing sector, dairy processing is among the industries with the highest energy demand and is generally quite important economically in numerous regions across the globe. The dairy processing industry in the United States produced over $90 billion in product exports in 2010 and the cost of energy required is over $721 million on electricity and $542 million on fuel [5]. A significant portion of the thermal energy used by the dairy industries, roughly 70%, is needed for process heating, which occurs at temperatures between 50°C and 250°C as shown in Figure 5.3. It has been stated that the energy consumption per kilogram of various dairy products might vary from 2.7 to 50 MJ. This portion of energy is provided by biomass in the boilers together with substantial assistance from fossil fuels. Heat is mostly needed during the dairy production process for pasteurisation and boiler preheating. Pasteurisation is a method that involves heating raw milk below its boiling point in order to eliminate or deactivate bacterial pathogens. For dairy products to have a longer shelf life, this kind of treatment is necessary. The manufacturing of milk powder involves numerous thermal processes and accounts for 15% of the total energy consumption in the dairy sector [6].

5.2.2 Bakery Industry

The bakery sector is one of the most important ones in the food companies. The predicted value of the worldwide bakery market in 2020 was $216 billion. The steps involved in producing bread, cake, cookies, nougat, brownies, and other baked goods include combining and kneading ingredients according to a recipe to create a homogenous dough that will then go through shaping and baking procedures. Equipment related to energy consumption is needed to produce these items, including the procedures used in production, including mixing or kneading, leavening, baking; according to a case study conducted in Quito, Ecuador, the energy indicator associated with processes in bakery industry is shown in Figure 5.4 [8].

The amount of thermal energy consumed for baking in facilities that use both fuel and electricity normally ranges from 2.4 to 6.9 MJ/kg [9]. Each country's industrial bakery has a distinct share of the market; in the UK, it accounts for 80% of the market, whereas in Germany, France, and Spain, it is 40%, 30%, and 19%, respectively [8] (Figure 5.5).

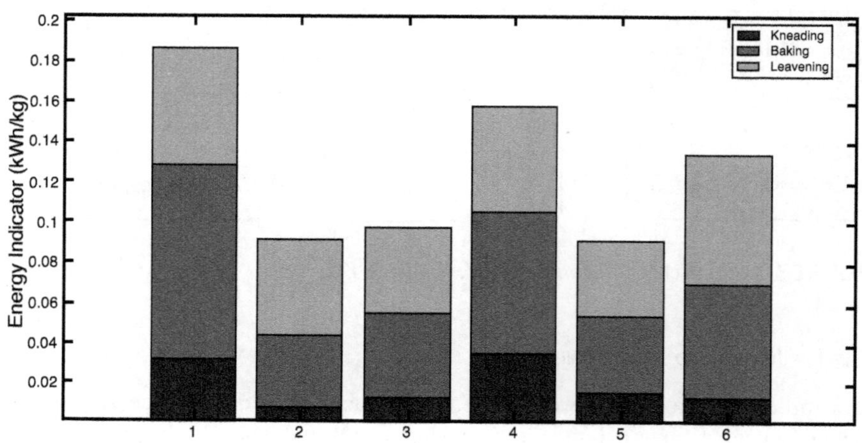

FIGURE 5.4 Energy indicator used by major bakery processes [8].

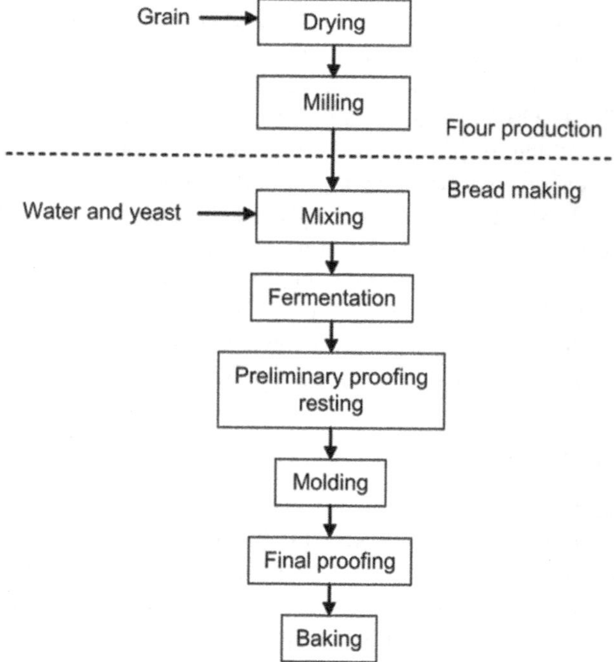

FIGURE 5.5 Flow chart showing the key procedures for producing flour and baking bread [9].

There are variations in energy indicators among the various food-producing industries. The largest energy indication in the USA was found in the preparation and packing of seafood products, whereas the processing of dairy products and animal feed had relatively low energy indicators [10].

5.2.3 Fruit and Vegetable Processing

Fruits and vegetables go through a number of active biological processes after they have been harvested, including respiration, ripening, and senescence. These processes can result in major quality changes in some fruits and vegetables, hence postharvest storage conditions and processing procedures must be carefully managed to avoid these changes [11]. Depending on the properties of the crops and the finished goods, grading, washing, chilling, and peeling are usually the first steps in the basic processing of fruits and vegetables. Before extracting juice, citrus fruits are typically graded and cleaned in a juice-processing facility. The juice is extracted as soon as possible from the peel, membrane, and seeds to preserve quality. The finishing procedure accomplishes additional pulp removal. According to the existing literature, the process of extraction necessitates a very little quantity of energy, around 0.03 MJ of electricity per kilogram of final juice. On the other hand, refining or finishing consumes about one-third of that energy [9]. Different fruits and vegetables have different extraction processes, which are often process-specific (Figure 5.6).

5.2.4 Sugar and Confectionary Processing

The various production processes used in candy manufacturing are caramel cooking, rolling, granulating, drying, and chilling which determine the energy consumption structure of the facility. The production of cocoa, chocolate, and confections (apart from bakery items) makes up one of the top 15 most energy-intensive food industry subsectors in Germany [12] (Figure 5.7).

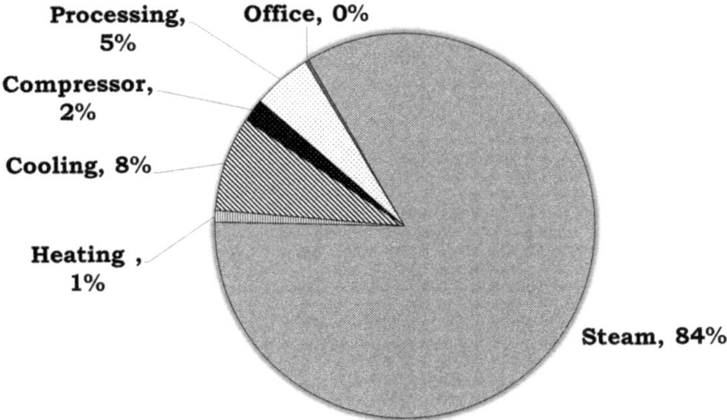

FIGURE 5.6 Typical energy use in the fruit juice industry [3].

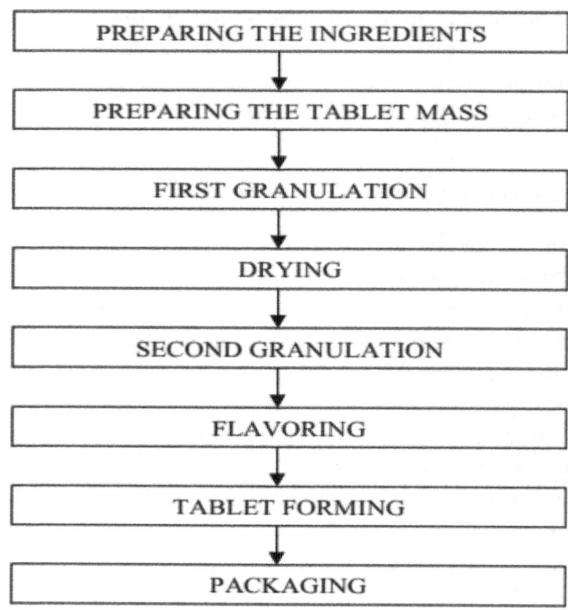

FIGURE 5.7 Tablet candy production process [12].

5.2.5 Meat, Poultry, and Fish Processing

Because meat processing plants produce a wide range of final products, processing energy needs are also very variable. The meat may also be deboned, chopped, ground, cooked, dried, smoked, chilled, or frozen at the facility as shown in Figure 5.8. For every kilogram of dressed carcass, cutting and deboning may need an extra 0.76 MJ of primary energy. One kilogram of ground beef needs an extra 0.3 and 5.1 MJ of power to grind and fry in order to make hamburgers.

The additional energy needed to freeze meat after initial processing varies depending on the type of carcass (in primary energy): 0.72 MJ for beef, veal, or sheep, 1.0 MJ for pig, and 1.2–2.4 MJ for poultry [9]. As compared to other food industries, the animal slaughtering sector uses the most electricity, while the sugar processing business uses the least [14].

5.2.6 Grain and Oilseed Milling

The oil can be extracted by milling the ground material (seed) followed by sorting and cleaning, removing husks, and crushing. In order to aid extraction, the milled plant material may additionally undergo conditioning (heating) prior to extraction [9]. Each type of oil can be processed in a broad variety of ways, which means that energy usage can also vary greatly. In a particular palm oil extraction mill, the various processes involved are threshing, boiling, pressing, and oil drying. Table displays the amount of energy (MJ/kg) used in a fully automated palm oil mill in Ghana to process 1 kg of palm fruit [15].

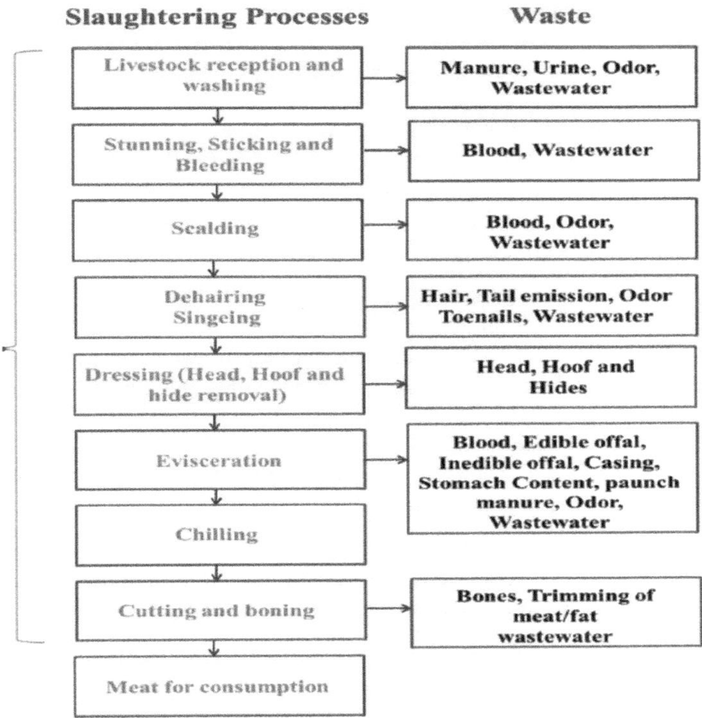

FIGURE 5.8 Flowchart showing the process of slaughter and garbage production [13].

Unit/Operation	Electrical Energy	Thermal Energy
Threshing/stripping	0.0048	0
Boiling	0	0.0292
Pressing/expulsion	0.0213	0
Oil drying	0	0.0225

5.3 PROCESSES OR END-USES TECHNOLOGIES IN FOOD INDUSTRY

Here are some of the end-use processes being used in most of the processing industries and require significant amount of energy for their operation.

5.3.1 Cleaning and Sanitation

Ensuring food products are not contaminated and upholding hygiene standards necessitate proper cleaning, which can include significant water, chemical, and energy use. During equipment cleansing procedures, energy is needed to generate hot water as well as steam.

5.3.2 Drying and Dehydration

In the food sector, the processes of evaporation and drying are typically utilised to concentrate and dry food items, respectively. These procedures make it possible to lower the water activity, or availability, and prevent the growth of microorganisms while in storage. There is a difference between the processes of dehydration and drying. It is actually true that a food product is considered dehydrated when it includes more than 2.5% water (d.b.), and called dry when the water content is below it [16]. One of the most crucial stages in moisture control for food product preservation and value addition is dehydration. The fundamental methods by which water is made to evaporate and the ensuing vapour is extracted either spontaneously or with coercion, leading to dehydration, are nothing but distinct heat transfer mechanisms. For most fresh products, the dehydration process alone accounts for up to 30% of the total processing cost [17].

5.3.3 Baking

Electric ovens with a maximum temperature of 250°C are employed at this step; normal cooking temperatures range from 190°C to 250°C. This will take place for a duration of 20–45 minutes, depending on the size and amount of the item to be prepared [2].

5.3.4 Heating and Cooking

Heat is necessary for several food processing procedures, including pasteurisation, sterilisation, and cooking. Depending on the particular needs of the process, this heat may be produced using steam, electricity, natural gas, or other sources.

5.3.5 Refrigeration and Freezing

Refrigeration is heavily utilised in food processing plants. Up to 15% of the world's total energy consumption is thought to be used by the refrigeration system [18]. One of the major issues facing the food business is the efficient management of bacterial activity. Apart from the previously mentioned techniques, the product's temperature can also be lowered by freezing it, which lowers its water activity and prevents the growth of germs, or by chilling it below the ideal growth temperature. The freezing process uses a significant amount of energy [16]. All sectors have similar needs for lighting and heating, ventilation, and air conditioning (HVAC); however, the sugar industry only makes up a very small portion of these needs [14].

5.3.6 Milling and Grinding

Energy is required to run milling and grinding machinery, which reduces raw materials into finer particles or powders, in order to produce a variety of food products, including flour, spices, and cereal grains. The operations of solvent extraction, mechanical pressing, and refining account for the majority of energy use in oilseed milling. In mechanical extraction, the oil is essentially forced out of the seeds by compressing them between barriers. The most popular mechanical extraction techniques include the use of screw or hydraulic presses [19].

5.3.7 Packaging

In general, the processing, packaging, distribution, and preparation of food consume 40% of the fossil energy used to produce it. In order to guarantee food security, protection, and waste avoidance, packaging is essential. Energy consumption can vary based on the type of packaging material used. Recycling packaging and using other waste package management techniques also use energy; therefore, these factors must be accounted [20].

5.3.8 Storage

In order to preserve food quality and deliver both safe and high-quality meals, this method involves maintaining food at the appropriate temperature. Refrigerators and freezers are common examples of energy-intensive equipment used for storage in the food business. A retail food product's storage requires between 1 and 3 MJ/kg [2].

5.4 ENERGY SOURCES USED IN FOOD INDUSTRIES

Food processing industries typically utilise a mix of energy sources, including electricity, natural gas, biomass, and in some cases, renewable energy sources like solar or wind power. Typically, electricity provides 15% of the energy required by the food sector. The majority of electrically powered processes are not able to replace other energy sources. The manufacturing sectors utilised 7% of the total electricity, which was consumed by the food industry. Fossil fuels are also utilised to generate energy, with natural gas being the most common type. The industries with the highest electricity and fuel costs were meat, grain mill products, and preserved fruits and vegetables [21]. The prime energy sources employed in food processing industries are coal, electricity, natural gas, petroleum, and renewable energy sources. The need for steam is high in the majority of food processing facilities. Approximately one-third of the energy used for fuel and electricity in the food sector, and more than half (57%) of the fuel used in the food processing sector, is used in boilers, which produce steam for supplying process heat to various unit operations like pasteurisation, evaporation, dehydration, and sterilisation [10]. Since energy use has a major impact on both operating costs and environmental sustainability, it is a crucial component of operations in the food processing sectors. About 200 EJ are consumed by the food industry annually worldwide, with processing and distribution accounting for 45% of the total [22]. Given the continued growth of the world's population, it is anticipated that by 2050, the demand for food will have increased by 60%, and the food sector would likewise require more energy [23]. Among the food production processes, food processing uses the greatest energy. The energy required to prepare food is thought to be three to four times higher on average than the energy required for primary production.

The amount of energy used by the food business varies based on the kind of the product and the country [24] as shown in Figure 5.9. Thermal heat generation uses

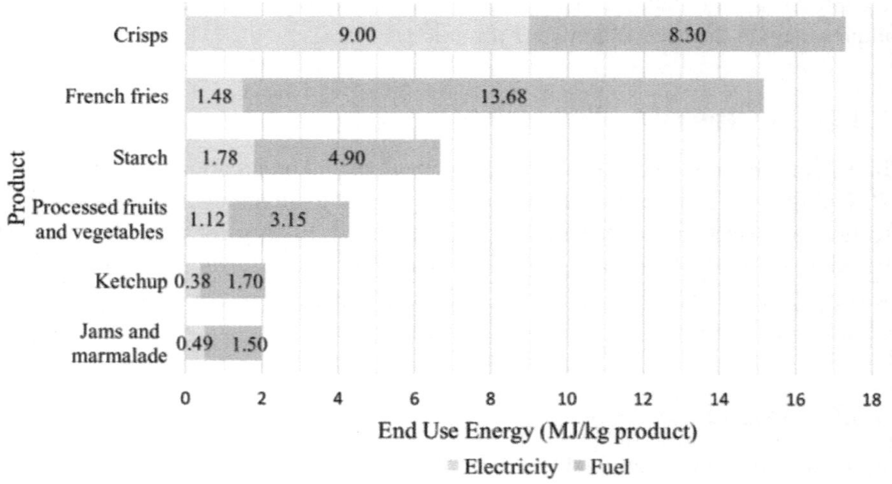

FIGURE 5.9 Energy consumption of some processed food items [22].

about 13% of the energy used. The food processing sector uses around 68% of its fuel for space and process heating and uses 16% for power motors, pumps, and other equipment; 6% is used for refrigeration; and 2% is used for compressors [25]. Of all the energy used in the food sector, 75% is for process heating and refrigeration, which includes ovens, furnaces, and refrigeration equipment. Motors, air compressors, fans, grinders, mixers, and other mechanical equipment consume 12%. About 8% of all expenses are related to infrastructure, such as building HVAC and lighting [14].

5.5 FACTORS AFFECTING ENERGY CONSUMPTION IN DIFFERENT STAGES OF FOOD PROCESSING

The energy consumption in food processing industries is governed by various parameters like the thermophysical properties of raw material, the technology used, equipment, range of processing temperatures, production capacity, and production process organisation [12]. The type of product has a major impact on how much energy is needed to make a particular quantity of it. As observed before, the energy needed to create per tonne quantity of an item is entirely different as compared to others (Table 5.1).

Factors responsible for energy consumption are provided below:

1. **Climate Conditions:** Due to the constant changes in climatic conditions, maintaining temperature control has become a significant challenge in food industries. This is especially true for food storage, where refrigeration and freezing demand considerable amounts of energy.
2. **Logistics Network:** Movement of goods from one location to another location requires large amount of energy. An optimised route and efficient supply chain is a necessary requirement for any type of industry to reduce the overall operational cost.

TABLE 5.1
Details of Applications Where Energy is Required

Stages of Food Processing	Sectors	Applications
Raw material production	Agriculture practices	• Irrigation • Fertilisation • Harvesting of agricultural products • Threshing
	Transportation	• Transportation of agricultural products from farms to processing industries
Processing of raw materials	Mechanical operations	• Crushing or grinding • Mixing • Winnowing
Manufacturing	Heat treatment	• Baking • Pasteurisation • Refrigeration
	Water treatment	• Water heating for cleaning of processed materials
	Packaging	• Packaging of processed materials
Storage and distribution	Freezing	• Keeping the processed material at low temperature
	Warehousing	• Lighting • Climate control

3. **Category of Food:** Energy consumption is totally based upon the type of food to be produced. For example, milk and dairy products consume more energy since it involves various thermal processes.
4. **Food Processing Techniques:** Food products which require high temperature for its production consumes more energy. For example, processing methods which involve bottling, frying, baking, and freezing require more energy.
5. **Size of Operation:** Mass production of food products consumes more energy. For example, companies like Nestle, Heinz, and Tip-Top utilise high machinery equipment for mass production which consumes more energy.
6. **Packaging:** Different types of packaging are available for various categories of food, each with its own energy consumption requirements.
7. **Economic Factors:** Public demand is also one of the main factors responsible for energy consumption. Both demand and supply are interlinked. If demand is more, then supply will be more, which overall affect the energy consumption of industry.

5.6 HEAT RECOVERY SYSTEMS

Installing heat recovery systems allows food processing facilities to capture and reuse excessive heat generated during several operations. This helps to improve overall energy efficiency by utilising waste heat for other heating purposes, such as preheating water or air. There are a number of sectors in food industry, which have potential sources of heat [26]. These include:

- Air compressors
- Boilers
- Gas turbines
- Refrigeration plant
- Evaporation
- Drying
- Pasteurisers
- Baking ovens
- Process heating
- Process cooling
- Distillation
- Sterilisation

Many technologies are available in the market to recover waste heat from industry like heat exchangers, regenerators, recuperators, and economisers, which are highly efficient to capture, recover, and heat exchange with some high heat transfer fluid. Some of the heat recovery systems for any specific food industry are mentioned below.

5.6.1 Plate Type Heat Exchanger

Plate type of heat exchanger is often used in the milk and dairy product industry. Here, the milk product requires process heating followed by process cooling simultaneously. So, to reduce energy consumption for both processes separately, plate type of heat exchanger is being used. In the plate type, heat exchanger number of metal plates is attached in an array form. These plates are being used to exchange between two fluids (Figure 5.10).

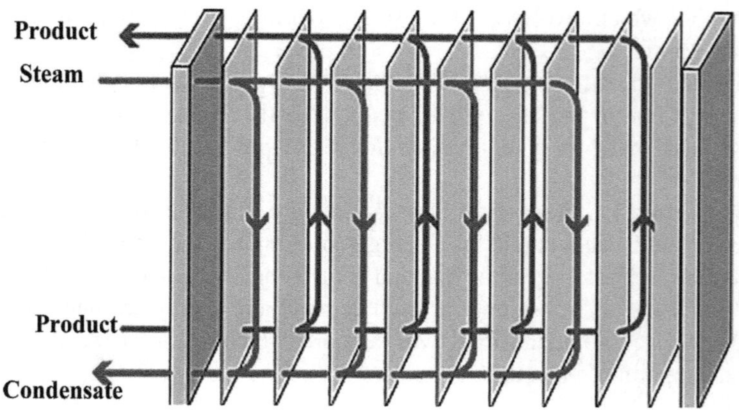

FIGURE 5.10 Plate type heat exchanger [27].

5.6.2 Heat Pipes

A heat pipe is made up of a metal tube that consists of a wick within the tube, which is filled with a specific volume of working fluid. The tube's evaporator and condenser are located at opposite ends. When the tube's evaporator end is heated, the liquid inside the tube evaporates, and the vapour moves to the other end of the tube. Vapour condenses and releases its latent heat at the opposite end, which is the condenser. To finish a cycle, the liquid uses capillary action to return to the beginning of the wick. Heat pipes are extremely effective because they transport heat from one place to another using latent heat rather than sensible heat [10].

5.6.3 Continuous Pasteuriser

The cold product is usually warmed by the heat used for pasteurisation in order to conserve energy. After pasteurisation, the hot product is utilised to warm up the cold product in a counter-current circulation system. This helps to save energy by precooling the pasteurised product and preheating the unpasteurised product. The procedure is carried out in a heat exchanger known as regenerative heat exchange [16] (Figure 5.11).

5.6.4 Organic Rankine Cycle

The organic Rankine cycle functions similarly to the steam-based Rankine cycle, but it uses organic fluids with low boiling temperatures and high vapour pressures to produce energy instead of water or steam. Experimental proof suggests that the system's utilisation of low-grade waste heat and power generation from geothermal, biomass, and solar energy sources is made possible by the use of an organic working fluid [27].

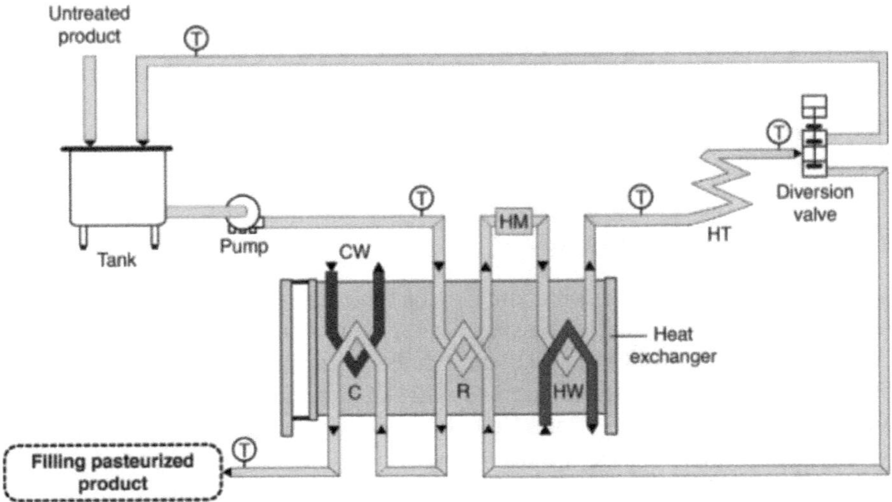

FIGURE 5.11 Working of continuous pasteuriser [16].

5.7 INNOVATIONS IN ENERGY-INTENSIVE OPERATIONS

5.7.1 Advanced Thermal Processing Technologies

Modern thermal processing technologies have many advantages over conventional methods, such as higher product quality, shorter processing times, and increased energy efficiency. By implementing these revolutionary techniques, the food sector can produce high-quality, minimally processed foods while reducing its energy use and environmental effect.

5.7.1.1 Microwave and Radio Frequency Heating

These methods sterilise food by using electromagnetic radiation. In general, microwaves produce more energy than other thermal treatments. Higher extraction yields are obtained when plant materials (such as oilseeds) are pretreated with microwave. Shorter cooking times were found to be a benefit of microwave cooking over other cooking techniques [16]. An innovative alternate drying technique that yields items of respectable quality is the combination microwave–hot air spouted bed (CMHS) method. When compared to materials dried using hot air and microwave drying processes, it allows for a significantly improved material quality and shorter drying time [28]. Another form of using electromagnetic waves is called radio frequency heating, which is a direct heating method that produces heat inside the product. Since the wave's penetration is deeper than that of microwaves, this method is more controllable [2]. In this technique, the frequency range is lower than that of microwave heating. Currently, it is used in industry for meat processing, drying, thawing, baking, and blanching.

5.7.1.2 Ohmic Heating

In this method, electric current is allowed to flow directly through the beverage to produce heat; it is because of the material's conductivity that there won't be any losses, which is advantageous in comparison to other techniques. The conductivity of the liquid and the voltage differential in the field determine how much heat is produced [2]. Using this method to cook meat products made them cook faster, use less energy, and make safer products. Compared to ordinary cooking, this cooking yields a firmer sample. Reduced cooking loss and increased juiciness were the outcomes of Ohmic heating. Several studies have shown that Ohmic heating, alone or in combination with regular cooking methods, may yield safer beef products [29].

5.7.2 Non-Thermal Processing Technologies

Technologies that function well at ambient or sublethal temperatures are frequently referred to as "non-thermal processing." Microorganisms can be inactivated to varied degrees by high hydrostatic pressure, oscillating magnetic fields, pulsed electric fields, high-intensity ultrasound, ultraviolet light, and pulsed light [30]. Usually, these non-thermal processing durations are quite brief. For example, foods are subjected to pressures as high as 1,000 MPa for a brief period of time during high-pressure processing. The basis of pulsed electric field treatment is the application of high-intensity electric field pulses for a few milliseconds, ranging from 5 to 55 kV/cm [18].

5.7.3 Efficient Refrigeration and Cooling Systems

In refrigerating plants, energy required is mainly in the form of compressor work. When compared to systems operating at a constant speed, variable speed drives drastically reduce energy usage by adjusting the motor speed in response to the cooling requirement. Similarly, use of modern insulating materials is essential for improving the cooling and refrigeration systems' effectiveness in the food processing industry. These materials assist in maintaining ideal temperatures by lowering heat gain/loss, which enhances energy efficiency and lowers operating expenses. Analysis was done on novel refrigeration cycles in small units, implication of adsorption system which is a form of mechanical refrigeration machine that eliminates the necessity of compressor and can be operated using low-grade fuels as well as renewable energy sources.

5.7.4 Renewable Energy Integration

It is anticipated that renewable energy sources will contribute more to the overall production of electricity. As a result, the food industry was also encouraged to use renewable energy for the supply of energy loads [2]. Due to constraints like limited availability and environmental challenges associated with conventional energy sources, renewable energy sources possess significant potential compared to traditional options. Wind, biomass, solar radiation (thermal or photovoltaic [PV]), and ground heat are the most attractive renewable energy sources for applications in industry.

5.7.4.1 Solar Thermal Systems

Thermal processes may benefit from solar energy as well; the direct use of solar energy is made for drying agricultural products in rural areas and commonly used for water heating applications these days. The two most well-known solar technologies are PV and solar thermal. Solar thermal systems capture solar energy through the use of collector technology. It was estimated that by 2020, about 95 SIPH plants with a combined installed capacity of 41 MW and a collector area of 69,124 m^2 will be operational worldwide in the food industry. It can be seen in figure that among these plants, 38% have Flat plate collector (FPC) technology installed, while 20% have Parabolic trough collector (PTC) and 20% have ETC [31,32] (Figure 5.12).

5.7.5 Innovative Drying Technologies

5.7.5.1 Freeze Drying

A highly effective preservation technique that keeps food quality, nutritional content, and structure intact is freeze drying, which involves crystallising the solvent (typically water) and/or suspension medium at a low temperature, followed by a direct sublimation from the solid state into the vapour phase. With only 1% to 4% moisture in the finished product, a long shelf life its advantages make it a preferred option in the food processing, pharmaceutical, and biotechnology industries, despite its higher cost and energy requirements. Freeze-dried products satisfy the needs of

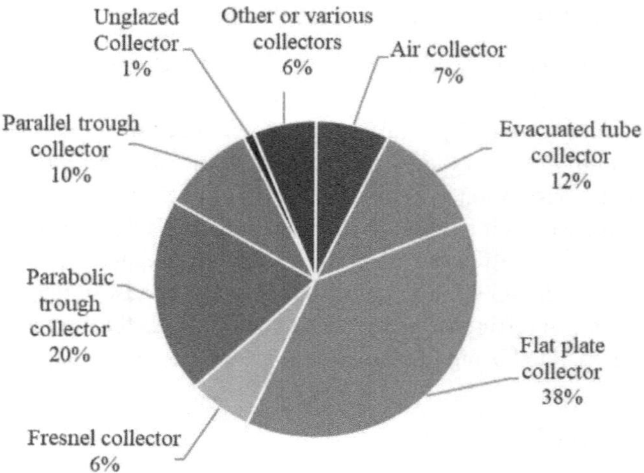

FIGURE 5.12 Types of solar collectors installed in the food industry [31].

both customers and industries for high-quality, long-lasting products by providing an extended shelf life and simplicity of rehydration [33].

5.7.5.2 Microwave Vacuum Drying (MVD)

Combining the benefits of vacuum drying with microwave heating, MVD is a promising process for producing high-quality dried foods. Higher energy efficiency and better product quality are guaranteed by the MVD's low temperature and quick mass transfer. For MVD, the cost of electric energy per unit was just 6.9% and around 68.8% of the costs associated with freeze drying and hot air drying, respectively. For this reason, MVD can be utilised as an excellent and affordable substitute for dehydrating food [34].

5.7.5.3 Ultrasonic-Assisted Drying (USN)

It is an advanced drying technique that uses high-frequency ultrasonic waves to enhance the drying process. Ultrasounds have been used in several operations, including cutting, drying, sterilising, and the extraction of bioactive substances, among many others. The biggest advantage of using USN in conjunction with air drying is that as compared to air drying alone, it can drastically reduce the drying temperature and/or drying time. For instance, it was demonstrated that the drying periods for carrot slices were reduced from 35 to 25 minutes by using air drying alone at 60°C and air drying in conjunction with USN at the same temperature [16].

5.7.6 Waste Management in Food Processing

Worldwide, 11.2 billion tonnes of solid garbage is expected to be produced annually; by 2025, it is projected to rise to 19 billion tonnes [13]. Utilising supposedly inedible byproducts in a biogas plant is a real-world example of reducing the expense of processing slaughterhouse waste. Manufacturing fat, meat and bone meal, and soil enhancers was the most prevalent way that industrial meat processing waste was disposed of. Utilising slaughterhouse waste as a raw material for the manufacture of biogas was one of the alternative means of using the trash, which had a high carbon content [35]. In addition to postharvest losses brought on by inadequate storage space, preparing and packaging vegetables in accordance with consumer requirements also contribute significantly to waste production. The rotting, peeled, shelled, and scraped parts of vegetables or slurries are considered vegetable wastes. These wastes can be composted or processed for the production of biofuels, such as through controlled fermentation [36].

5.7.7 Energy-Efficient Packaging

Packaging items in a way that minimises the amount of energy needed for processing, storage, or transportation is known as energy-efficient packaging. Utilising energy-efficient production methods or cutting the number of processing steps can help design packaging that uses less energy. These methods include (i) optimising the packaging layout to lessen the requirement for extreme climate control in the holding spaces, (ii) lowering the energy needed for heating or refrigeration by maintaining temperature-sensitive goods with insulating materials. (iii) choosing recyclable or reusable materials to reduce energy use in manufacturing and waste processing and (iv) spreading weight evenly to increase transportation's fuel efficiency. In addition to cutting operating expenses, energy-efficient packaging promotes environmental sustainability by preserving natural resources and reducing greenhouse gas emissions.

5.8 CONCLUSION

The consumers might not always desire to eat in an energy-efficient manner. To ensure food quality and save the most energy during cooking, it is crucial to research energy-efficient cooking methods. Food industries possess substantial greenhouse gas emissions as well as energy consumption. It is vital to evaluate the forthcoming issues, given that the anticipated rise in food consumption is not expected to coincide with the expansion of energy supplies. Energy efficiency and sustainability are significantly improving as a result of innovations in energy-intensive operations in the food sector. While conventional and novel approaches to food processing energy efficiency seem to offer possibilities for the food processing industry to become carbon neutral, there is currently a lack of real implementation. In order to more effectively motivate users to implement energy-efficient practices, appropriate policies must be developed.

REFERENCES

1. A. Shukla and R. Kumar, "Food Processing Industry in India: Challenges and Potential." Available: https://www.researchgate.net/publication/339940192.
2. J. M. Clairand, M. Briceno-Leon, G. Escriva-Escriva, and A. M. Pantaleo, "Review of energy efficiency technologies in the food industry: trends, barriers, and opportunities," *IEEE Access*, 8, 48015–48029, 2020. doi: 10.1109/ACCESS.2020.2979077.

3. S. Naresh Kumar and B. Chakabarti, "Energy and carbon footprint of food industry," In: Muthu, S. (eds) *Environmental Footprints and Eco-Design of Products and Processes*, Springer, Singapore, 2019, pp. 19–44. doi: 10.1007/978-981-13-2956-2_2.
4. T. P. De Araújo, M. M. De Moraes, C. Afonso, C. Santos, and S. S. P. Rodrigues, "Food processing: comparison of different food classification systems," *Nutrients*, 14(4), 14040729, 2022, doi: 10.3390/nu14040729.
5. R. Briam, M. E. Walker, and E. Masanet, "A comparison of product-based energy intensity metrics for cheese and whey processing," *Journal of Food Engineering*, 151, 25–33, 2015, doi: 10.1016/j.jfoodeng.2014.11.011.
6. S. N. Moejes and A. J. B. van Boxtel, "Energy saving potential of emerging technologies in milk powder production," *Trends in Food Science and Technology*, 60, 31–42, 2017. doi: 10.1016/j.tifs.2016.10.023.
7. A. K. Sharma, C. Sharma, S. C. Mullick, and T. C. Kandpal, "Potential of solar industrial process heating in dairy industry in India and consequent carbon mitigation," *Journal of Cleaner Production*, 140, 714–724, 2017, doi: 10.1016/j.jclepro.2016.07.157.
8. M. Briceño-León, D. Pazmiño-Quishpe, J. M. Clairand, and G. Escrivá-Escrivá, "Energy efficiency measures in bakeries toward competitiveness and sustainability – case studies in Quito, Ecuador," *Sustainability (Switzerland)*, 13(9), 13095209, 2021, doi: 10.3390/su13095209.
9. R. O. Morawicki and T. Hager, "Energy and greenhouse gases footprint of food processing," In *Encyclopedia of Agriculture and Food Systems*, Elsevier, Amsterdam, The Netherlands, 2014, pp. 82–99. doi: 10.1016/B978-0-444-52512-3.00057-7.
10. L. Wang, "Energy efficiency technologies for sustainable food processing," *Energy Efficiency*, 7(5), 791–810, 2014, doi: 10.1007/s12053-014-9256–8.
11. N. Sumonsiri and S. A. Barringer, "*6 Fruits and Vegetables-Processing Technologies and Applications*," Wiley, Hoboken, NJ, 2014.
12. J. Wojdalski et al., "Energy efficiency of a confectionery plant – case study," *Journal of Food Engineering*, 146, 182–191, 2015, doi: 10.1016/j.jfoodeng.2014.08.019.
13. V. Mozhiarasi and T. S. Natarajan, "Slaughterhouse and poultry wastes: management practices, feedstocks for renewable energy production, and recovery of value added products," *Biomass Conversion and Biorefinery*, 2022. doi: 10.1007/s13399-022-02352-0.
14. M. Compton, S. Willis, B. Rezaie, and K. Humes, "Food processing industry energy and water consumption in the Pacific northwest," *Innovative Food Science and Emerging Technologies*, 47, 371–383, 2018. doi: 10.1016/j.ifset.2018.04.001.
15. G. A. Akolgo et al., "Energy analysis for efficient mechanisation of palm oil extraction in Ghana: Targeting circular economy," *Energy Reports*, 10, 4800–4807, doi: 10.1016/j.egyr.2023.11.018.
16. S. Roohinejad, R. Greiner, O. Parniakov, M. Koubaa, and N. Nikmaram, "Energy saving food processing," In *Sustainable Food Systems from Agriculture to Industry: Improving Production and Processing*, Elsevier, Amsterdam, The Netherlands, 2018, pp. 191–243. doi: 10.1016/B978-0-12-811935-8.00006-8.
17. A. R. Eswara and M. Ramakrishnarao, "Solar energy in food processing – a critical appraisal," *Journal of Food Science and Technology*, 50(2), 209–227, 2013. doi: 10.1007/s13197-012-0739-3.
18. L. Wang, "*6 Energy Consumption and Reduction Strategies in Food Processing*." Wiley, Hoboken, NJ, 2013.
19. R. B. L. Temporim, A. Petrozzi, V. Coccia, F. Cotana, and G. Cavalaglio, "A prototype plant for oilseed extraction: analysis of mass and energy flows," *Sustainability (Switzerland)*, 12(22), 1–11, 2020, doi: 10.3390/su12229786.
20. P. Peduzzi and R. Harding Rohr Reis, "The end to cheap oil: a threat to food security and an incentive to reduce fossil fuels in agriculture," *Environmental Development*, 3, 157–165, 2012, doi: 10.1016/j.envdev.2012.05.008.

21. S. Drescher, N. Rao, J. Kozak, and M. Okos, "*A Review of Energy Use in the Food Industry*." Purdue University, West Lafayette, IN, 1997.
22. A. Ladha-Sabur, S. Bakalis, P. J. Fryer, and E. Lopez-Quiroga, "Mapping energy consumption in food manufacturing," *Trends in Food Science and Technology*, 86, 270–280, 2019, doi: 10.1016/j.tifs.2019.02.034.
23. F. M. Kanchiralla, N. Jalo, P. Thollander, M. Andersson, and S. Johnsson, "Energy use categorization with performance indicators for the food industry and a conceptual energy planning framework," *Applied Energy*, 304, 117788, 2021, doi: 10.1016/j.apenergy.2021.117788.
24. B. Bajan, J. Łukasiewicz, and A. Mrówczyńska-Kamińska, "Energy consumption and its structures in food production systems of the visegrad group countries compared with eu-15 countries," *Energies (Basel)*, 14(13), 14133945, 2021, doi: 10.3390/en14133945.
25. H. Wu, S. A. Tassou, T. G. Karayiannis, and H. Jouhara, "Analysis and simulation of continuous food frying processes," *Applied Thermal Engineering*, 53(2), 332–339, 2013, doi: 10.1016/j.applthermaleng.2012.04.023.
26. D. Reay, "Heat recovery in the food industry," In J. Klemeš, R. Smith, and J.-K. Kim (eds) *Handbook of Water and Energy Management in Food Processing*, Elsevier Inc., Amsterdam, The Netherlands, 2008, pp. 544–569. doi: 10.1533/9781845694678.4.544.
27. H. Jouhara, N. Khordehgah, S. Almahmoud, B. Delpech, A. Chauhan, and S. A. Tassou, "Waste heat recovery technologies and applications," *Thermal Science and Engineering Progress*, 6, 268–289, 2018. doi: 10.1016/j.tsep.2018.04.017.
28. W. Jindarat, S. Sungsoontorn, and P. Rattanadecho, "Analysis of energy consumption in a combined microwave-hot air spouted bed drying of biomaterial: coffee beans," *Experimental Heat Transfer*, 28(2), 107–124, 2015, doi: 10.1080/08916152.2013.821544.
29. P. B. Pathare and A. P. Roskilly, "Quality and energy evaluation in meat cooking," *Food Engineering Reviews*, 8(4), 435–447, 2016, doi: 10.1007/s12393-016-9143–5.
30. R. N. Pereira and A. A. Vicente, "Environmental impact of novel thermal and non-thermal technologies in food processing," *Food Research International*, 43(7), 1936–1943, 2010, doi: 10.1016/j.foodres.2009.09.013.
31. M. I. Ismail, N. A. Yunus, and H. Hashim, "Integration of solar heating systems for low-temperature heat demand in food processing industry – a review," *Renewable and Sustainable Energy Reviews*, 147, 111192, 2021, doi: 10.1016/j.rser.2021.111192.
32. P. K. Jangde, A. Singh, and T. V. Arjunan, "Efficient solar drying techniques : a review," *Environmental Science and Pollution Research*, 29(34), 50970–50983, 2021.
33. A. Ciurzyńska and A. Lenart, "Freeze-drying – application in food processing and biotechnology – a review," *Polish Journal of Food and Nutrition Sciences*, 61(3), 165–171, 2011, doi: 10.2478/v10222-011-0017-5.
34. N. Jiang, C. Liu, D. Li, and Y. Zhou, "Effect of blanching on the dielectric properties and microwave vacuum drying behavior of Agaricus bisporus slices," *Innovative Food Science and Emerging Technologies*, 30, 89–97, 2015, doi: 10.1016/j.ifset.2015.05.001.
35. S. Bielski, A. Zielińska-Chmielewska, and R. Marks-Bielska, "Use of environmental management systems and renewable energy sources in selected food processing enterprises in poland," *Energies (Basel)*, 14(11), 14113212, 2021, doi: 10.3390/en14113212.
36. A. Singh, A. Kuila, S. Adak, M. Bishai, and R. Banerjee, "Utilization of vegetable wastes for bioenergy generation," *Agricultural Research*, 1(3), 213–222, 2012, doi: 10.1007/s40003-012-0030-x.

6 Technological Innovations for a Sustainable Low-Carbon Future

S. Vijayan, V. Sureshkannan, R. Prakash, and M.M. Matheswaran

6.1 INTRODUCTION

Environmental protection is increasingly seen as essential for sustainable long-term development, not a constraint on economic growth. Eco-innovation's capacity to meet economic and environmental goals has garnered global attention. Technology drives eco-innovation, economic growth, and sustainability (Albino et al., 2014). The importance of low-carbon technology is extremely significant in the worldwide endeavour to address climate change. These technologies are essential for optimizing energy usage, mitigating greenhouse gas emissions, and encouraging the growth of sustainability. The adoption of low-carbon technologies has become essential in limiting the harmful effects of climate change. These innovative approaches help to mitigate the effects of extreme weather, decrease sea level rise, and conserve biodiversity, by decreasing the consumption of fossil fuels (IPCC, 2018). Moreover, they offer economic benefits, including job creation in emerging green industries and reduced health costs associated with air pollution (IRENA, 2020b).

Reducing carbon emissions is imperative for mitigating climate change, improving public health, and ensuring economic stability. High atmospheric concentrations of carbon dioxide (CO_2) and other greenhouse gases (GHGs) are the main causes of global warming, which causes serious environmental problems. The Intergovernmental Panel on Climate Change (IPCC, 2018) states that drastic changes in the energy, land, urban, and industrial sectors must happen quickly in order to keep global warming to 1.5°C. Annually, millions of premature fatalities are caused by air pollution caused by the burning of fossil fuels, which is associated with respiratory and cardiovascular diseases (World Health Organization, 2021). Transitioning to cleaner energy sources can substantially enhance public health outcomes and improve air quality by lowering carbon emissions. Furthermore, the economic impact of climate change is profound, with the potential to cause substantial financial losses due to damage to infrastructure, decreased agricultural productivity, and increased healthcare costs (Maréchal, 2007). In addition to reducing these risks, investments in energy-efficient and renewable energy technologies help stimulate the economy by generating jobs and advancing technology. Energy security is another critical aspect. Dependence on fossil fuels often leads to geopolitical tensions and economic volatility. Countries may improve

DOI: 10.1201/9781003452072-6

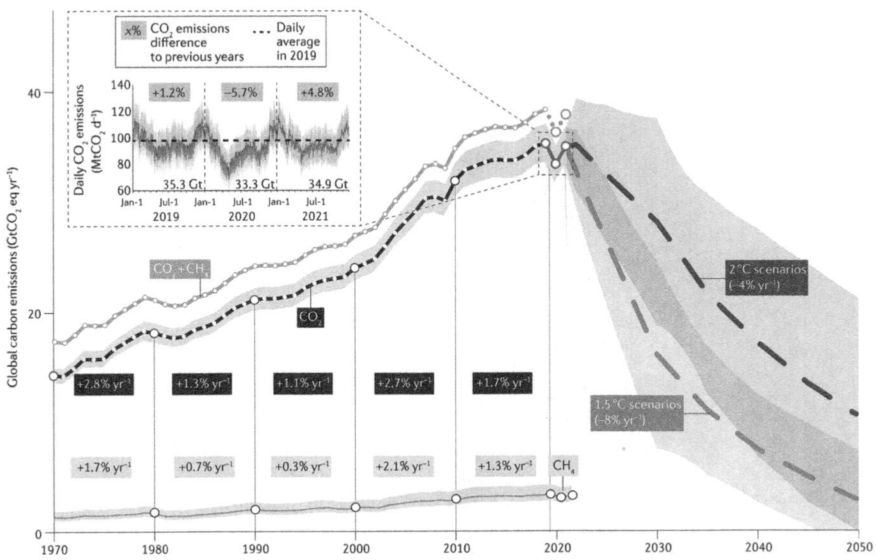

FIGURE 6.1 CO_2 emissions around the world from 1970 to 2021 (Liu et al., 2022).

energy security and prevent supply interruptions by adopting renewable energy (IEA, 2020). Additionally, climate change disproportionately affects low-income and vulnerable communities, exacerbating social inequities. Reducing carbon emissions can help protect these populations from the worst impacts of climate change. Global CO_2 emissions have shown a significant upward trend from 1970 to 2022, driven primarily by fossil fuel combustion and cement production (Figure 6.1). The COVID-19 pandemic caused a temporary decline in emissions due to reduced industrial activity and transportation, resulting in emissions dropping to 33.4 $GtCO_2$. Emissions rebounded postpandemic, surpassing prepandemic levels. By 2022, global CO_2 emissions reached 36.1 ± 0.3 GtCO2 (Liu et al., 2023). This period has been characterized by industrial growth, increased energy consumption, and a rise in global population, all contributing to higher carbon emissions.

This chapter explores a comprehensive summary of key technological innovations that contribute to a low-carbon environment. The objectives are to identify and analyse the possibility of these technologies in limiting carbon emissions, assess their current and future applications across various sectors, and highlight the challenges and needs for further innovation. It comprises renewable energy, energy storage, and carbon capture and utilization (CCU) methods, smart grid and energy management systems, and emerging materials and processes. Through this analysis, the survey seeks to communicate with industry stakeholders and policymakers about the advancements and opportunities in low-carbon technologies.

6.2 CARBON MITIGATION POTENTIAL

The potential for carbon mitigation through technological innovations is vast, offering significant opportunities to reduce greenhouse gas emissions across multiple sectors. With the ongoing increase in global temperatures and the growing severity of

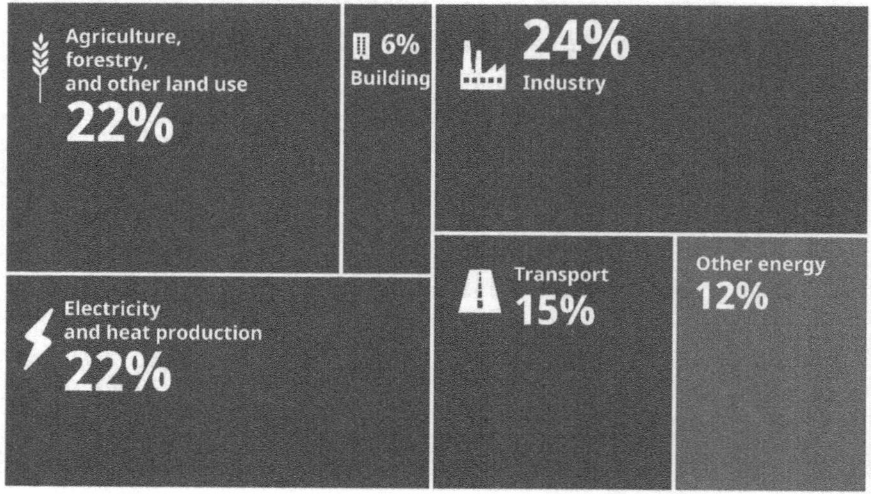

FIGURE 6.2 Global CO_2 emissions by economic sector in 2019 (IPCC, 2022).

climate change consequences, the importance of deploying effective carbon mitigation technologies becomes increasingly urgent (IPCC, 2018). This section explores various technologies and their possibility to contribute to a low-carbon future, focusing on renewable energy sources, energy storage solutions, and CCU.

Focusing on five key economic sectors—industry, electricity, agriculture, transport, and buildings—offers a strategic approach to addressing nearly 90% of GHG emissions (indicated in Figure 6.2). Governments can promote the transition to a low-carbon economy and facilitate sustainable development through the implementation of various policies and advancements in technology. By leveraging these policy levers across sectors, governments can drive systemic change toward a low-carbon future and pave the way for sustainable and resilient societies.

The attainment of net-zero emissions by 2050 necessitates a significant change in energy policies, infrastructure investments, and technological advancements across various sectors. The pathway to net zero emphasizes a sharp decline in fossil fuel reliance, a transition to clean electricity, and substantial investments in enabling technologies and infrastructure (International Energy Agency, 2021) are depicted in Figure 6.3.

Different sectors contribute variably to global carbon emissions, and thus, sector-specific mitigation strategies are essential. The energy sector is the primary producer of global emissions, comprising approximately 73% of the total (IEA, 2020). The main approaches for reducing emissions in this sector are shifting toward renewable energy sources and enhancing energy efficiency. Solar and wind power are expected to dominate future energy mixes, significantly reducing emissions from power generation (IRENA, 2019). The industrial sector is responsible for over 24% of worldwide emissions and offers substantial prospects for reducing emissions through

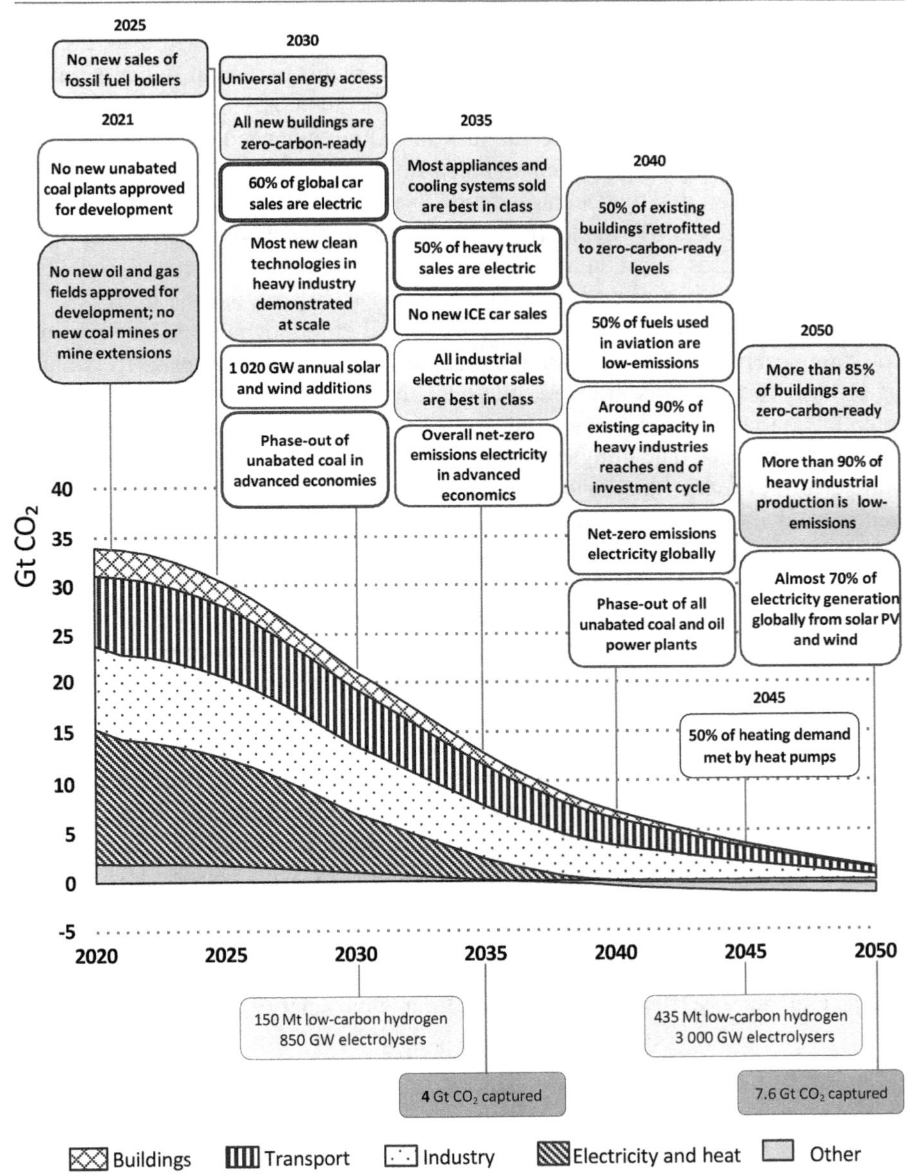

FIGURE 6.3 Key milestones in the pathway to net zero by 2050 (International Energy Agency, 2021).

enhancements in energy efficiency, transitioning to alternative fuels, and implementing CCS technology. Innovations in industrial operations, such as the use of hydrogen in steel production and the development of low-carbon cement, can further reduce emissions (IEA, 2020). The transportation sector, contributing around 15% of global emissions, can achieve substantial reductions through the use of electric vehicles (EVs), improvements in fuel efficiency, and the development of sustainable biofuels (IEA, 2020). Implementing policies that encourage the use of public transit, cycling, and walking can effectively decrease emissions from this particular sector (Creutzig et al., 2015). The buildings sector, responsible for about 6% of global emissions, can benefit from energy-saving techniques including enhanced insulation, energy-efficient lights, and HVAC systems (Pérez-Lombard et al., 2008). Implementing these measures can greatly decrease energy use and emissions from residential and commercial buildings. By deploying a combination of technological, behavioural, and policy strategies, substantial emission reductions can be achieved, contributing to the worldwide initiative to address climate change (Figures 6.4 and 6.5).

Figure 6.6 shows mitigation options, costs, and 2030 potentials. The horizontal axis shows each option's mitigation potential relative to a baseline scenario, comparing the net lifetime costs of decreasing GHG emissions to a reference technology. Potentials, target years, reference scenarios, and cost evaluations are defined using a consistent manner. The mitigation potential of each choice is shown by solid bars, while error bars show the range of outcomes across scenarios. Technical improvements, local implementation differences, and economies of scale affect cost estimates. This helps climate change mitigation policymakers and stakeholders assess mitigation strategy efficacy and cost.

Renewable energy sources, like solar, wind, and biomass, have shown a lot of promise for lowering carbon pollution and cutting our reliance on fossil fuels. Significant advancements have been made in the field of solar power recently, especially in solar photovoltaic (PV) technology. The efficiency improvements and cost reductions in PV cells have made solar energy more competitive with conventional energy sources (IRENA, 2020b). For example, the development of high-efficiency PV cells and innovative solar panel designs has increased the energy yield of solar installations, thereby enhancing their carbon mitigation potential (Shubbak, 2019).

Wind energy is another key renewable technology with considerable carbon mitigation potential. The installation of offshore wind farms has had the biggest effect because they get stronger and more regular winds than mainland installations. Innovations in turbine design and materials resulted in higher capacity factors and increased energy output, making wind energy a more reliable and

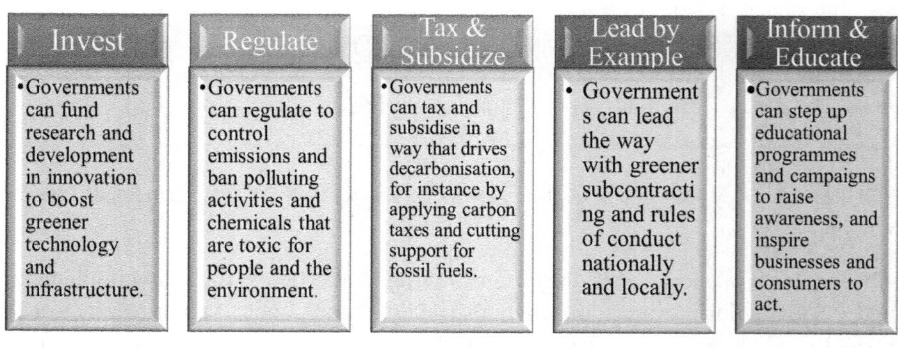

FIGURE 6.4 Government policies recommended for carbon emission mitigation.

Industry

- Emissions from heavy industries that produce primary materials like steel, chemicals, cement, glass and paper are energy intensive.
- Research and innovation, re-use and recycling, and improvement in manufacturing efficiency will need to be scaled up.

Electricity

- Strategies must emphasise decarbonising the electricity sector while providing affordable and reliable supply of electricity for all.
- Renewable technologies are already cost-competitive in an increasing number of countries.

Agriculture

- Focusing on agriculture and related land-use change will be critical to reduce emissions and can help replenish our forests and reverse biodiversity loss.
- There is also potential to use agricultural soils to capture more carbon.

Transport

- Decarbonising transport relies on innovative technologies and a shift towards cleaner alternatives.
- Integrated land-use and transport planning that reduces distances and encourages public transport, cycling and walking are crucial to this process.

Buildings

- Policies can improve building quality. Technologies and methods are much more climate friendly than before and often cheaper than conventional construction.
- Existing buildings can also become more energy efficient, through renovation and retrofitting.

FIGURE 6.5 Key economic sectors for carbon emission mitigation policy implementation.

significant contributor to the renewable energy mix (IRENA, 2020b). The scalability of wind energy projects, along with ongoing technological advancements, positions wind power as an important element in lowering global carbon emissions (Díaz-González et al., 2012).

Energy storage systems are necessary to stabilize renewable energy sources and reduce intermittency. Due to their energy density, efficiency, and low cost, lithium-ion batteries (LIBs) dominate energy storage. These batteries store extra energy from peak output to utilize during low production, improving renewable energy system stability and integration (Olabi, 2017). Additionally, advancements in solid-state batteries and other next-generation storage technologies promise further improvements in energy density, safety, and lifecycle, making them even more effective in supporting carbon mitigation efforts (Manthiram, 2017).

Hydrogen fuel cells represent another promising energy storage solution with significant carbon mitigation potential. Hydrogen can be produced from renewable energy sources through processes such as electrolysis, which splits water into hydrogen and oxygen using electricity. Hydrogen is a clean and sustainable energy source as it produces electricity in fuel cells with water as the sole byproduct. The versatility

Many options available now in all sectors are estimated to offer substantial potential to reduce net emissions by 2030. Relative potentials and costs will vary across countries and in the longer term compared to 2030.

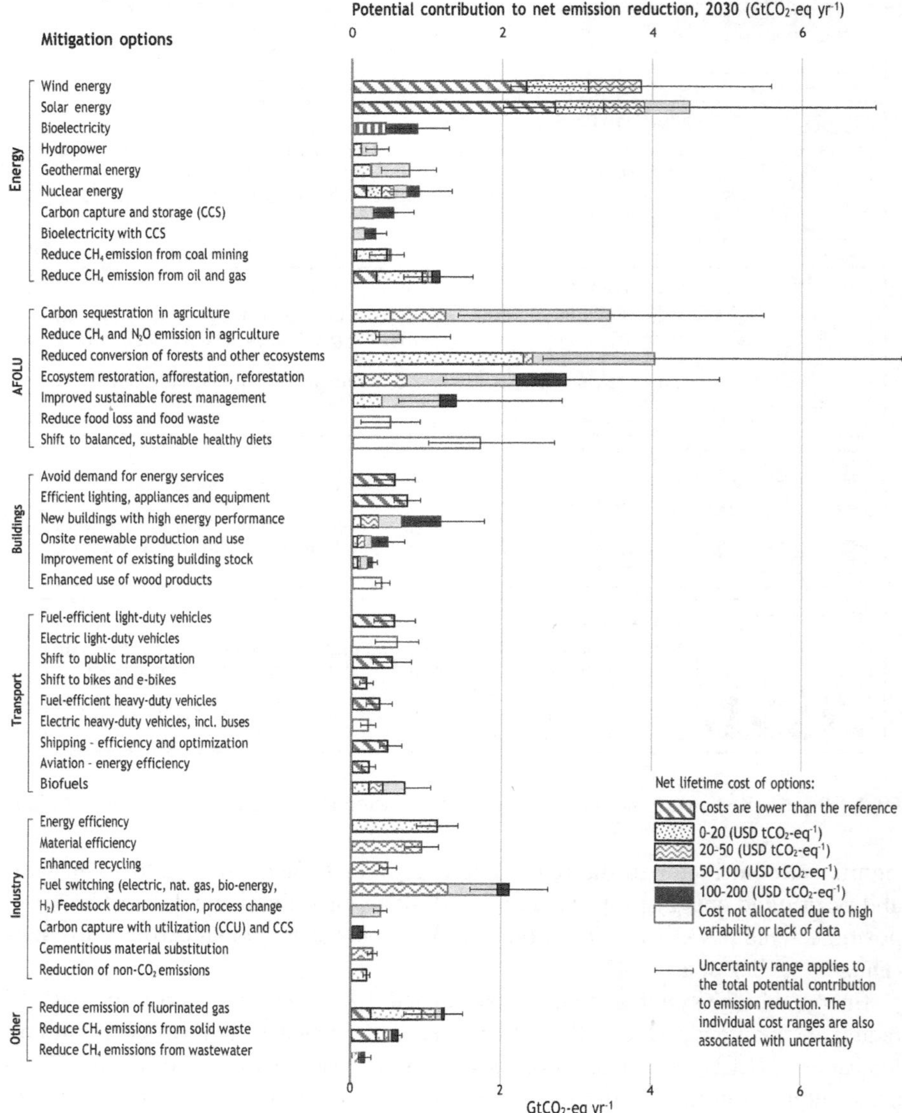

FIGURE 6.6 Relative potentials and costs for various mitigation options (IPCC, 2022).

of hydrogen fuel cells allows them to be used in various applications, including transportation, power generation, and industrial processes, thereby contributing to carbon emission reductions across multiple sectors (Staffell et al., 2019).

CCU technology aids in the difficult-to-decarbonize sectors including cement, steel, and chemical manufacture. Industrial carbon dioxide emissions are captured and stored underground or converted into fuels, chemicals, or construction materials by CCU.

Modern CCU technologies are more efficient and cost-effective, making them suitable for large-scale implementation (IEA, 2020). For instance, the introduction of new solvents, adsorbents, and membranes for CO_2 capture has enhanced the performance and reduced the energy requirements of CCU systems (Sanz-Pérez et al., 2016).

The incorporation of smart grid technologies and energy management systems further supports carbon mitigation by optimizing energy use and enhancing the efficiency of power distribution networks. Smart grids facilitate real-time monitoring and management of energy flows, allowing for better integration of renewable energy sources and improved demand response capabilities. Energy management systems, which utilize advanced software and data analytics, help industries and consumers optimize their energy consumption, reduce waste, and lower their carbon footprints (Fang et al., 2012).

6.2.1 Carbon Mitigation Strategies

Carbon mitigation strategies are essential in the battle against climate change, focusing on reducing GHG emissions and enhancing carbon sinks. These strategies can be broadly categorized into technological, behavioural, and policy approaches (IPCC, 2018). Technological strategies encompass the utilization of renewable energy sources, enhancements in energy efficiency, and the implementation of CCS technology. Renewable energy technologies, such as solar, wind, and biomass, replace carbon-intensive fossil fuels, significantly lowering carbon emissions (IRENA, 2020b). The reduction of energy consumption and emissions may be accomplished by the implementation of energy efficiency improvements in industrial processes, buildings, and transportation (Pérez-Lombard et al., 2008). Behavioural strategies involve changes in consumption patterns and lifestyle choices that reduce carbon footprints. This includes reducing energy consumption through conscious efforts, such as using public transportation, adopting energy-efficient appliances, and supporting sustainable products (Dietz et al., 2009). Public awareness campaigns and education can play a crucial role in promoting these behavioural changes. Policy initiatives facilitate technical and behavioural changes. These include carbon taxes and cap-and-trade systems, which generate economic incentives to reduce emissions (Stavins, 2008). Subsidies and incentives for renewable energy and energy efficiency projects, along with regulations and standards, can drive the adoption of low-carbon technologies (IRENA, 2020b).

6.2.2 Assessment of Global Carbon Mitigation Potential

Assessing the global carbon mitigation potential involves evaluating the cumulative impact of various technologies and strategies in decreasing GHG emissions. According to IRENA, renewable energy sources could account for approximately 90% of the required carbon reductions by 2050 to limit global temperature rise to 1.5°C (IRENA, 2019). Solar and wind energy, due to their vast availability and declining costs, are expected to play a leading role in this transition. Energy efficiency improvements also offer substantial mitigation potential. Building, industrial, and transportation energy efficiency improvements can cut emissions worldwide by up to 40% (IEA, 2020).

CCS technologies are critical for mitigating emissions from hard-to-decarbonize sectors such as cement, steel, and chemical manufacturing. The IEA estimates that CCS could contribute to approximately 15% of the global emission reductions needed by 2050 (IEA, 2020). However, the deployment of CCS is currently limited due to high costs and regulatory challenges, highlighting the need for continued research and development. Additionally, nature-based solutions, such as afforestation, reforestation, and soil carbon sequestration, can enhance carbon sinks and contribute to mitigation efforts. In addition to sequestering carbon, these tactics offer co-benefits like enhanced ecosystem services and biodiversity protection (Griscom et al., 2017).

6.3 TECHNOLOGY DESCRIPTION

6.3.1 Renewable Energy Technologies

Renewable energy generates electricity or heat from renewable sources. These technologies reduce carbon emissions and provide energy security by replacing fossil fuels. The following sections provide an overview of key renewable energy sources and their role in transitioning to a low-carbon future.

6.3.1.1 Solar Power Innovations

Solar power is the conversion of sunlight into energy via PV cells or concentrating solar power (CSP) devices. PV cells, which are often seen in solar panels, directly convert sunlight into energy via the PV effect. Recent developments in PV technology have focused on boosting efficiency, lowering prices, and increasing durability. Thin-film solar cells, for example, utilize lightweight and flexible materials, offering versatility in installation and application (Shubbak, 2019). Perovskite solar cells represent another promising innovation, demonstrating high efficiency and potential for low-cost production. CSP systems utilize mirrors or lenses to concentrate sunlight onto a small area, generating heat that drives a steam turbine or heat engine to produce electricity. CSP technologies have seen advancements in thermal energy storage (TES), enabling continuous electricity generation even after sunset or during cloudy periods (Gamil et al., 2022). Hybrid PV-CSP systems, combining PV and concentrating solar technologies, offer improved efficiency and grid integration (Singh et al., 2019). Overall, solar power innovations continue to drive down costs and expand the deployment of solar energy worldwide. PV technology, comprising solar trackers, inverters, and batteries, has become increasingly viable for diverse applications, thanks to advancements in materials and systems. It now powers homes, businesses, water pumps, satellites, and spacecraft, indicating its broad utility. As PV technology continues to advance, its adoption is expected to grow further across various sectors in the future (Dada & Popoola, 2023).

The advancement of bifacial solar cells, especially the interdigitated back contact (IBC) structure, represents a significant advancement in solar technology, offering unique capabilities to absorb light from both front and back surfaces, thereby increasing overall power generation compared to traditional monofacial cells. These high-efficiency architectures—such as IBC, passivated emitter and rear cell (PERC), heterojunction with intrinsic thin layer (HIT), and tunnel oxide passivated contact (TOPCon)—have achieved conversion efficiencies exceeding 25%, demonstrating their potential for widespread integration into various applications. To maximize these benefits, innovative integration methods like building integrated photovoltaics (BIPV), floating solar photovoltaic (FPV) systems, and agrivoltaics have been developed, each offering unique advantages such as aesthetic integration, efficient land use, and additional income streams. Optimization of panel orientation and tilt angle, alongside regular cleaning, are crucial for maintaining efficiency. Advanced monitoring and control systems, particularly those driven by artificial intelligence (AI), enhance these systems further. AI-based solar tracking optimizes panel orientation, AI-driven cleaning systems ensure consistent efficiency, and robust cybersecurity measures protect the infrastructure from cyber threats. Collectively, these advancements in solar cell technology, integration methods, and AI-driven enhancements are driving significant improvements in solar energy efficiency and sustainability, facilitating the shift for a cleaner future (Vodapally & Ali, 2022).

Solar energy is ideal for industrial thermal needs, reducing fossil fuel reliance and pollution. Low- and medium-temperature solar thermal collectors efficiently meet thermal energy needs in processes like cleaning, heat treatment, drying, and sterilization in various industries such as food processing, automotive, textile, and pharmaceuticals. These collectors offer significant potential for sustainable industrial processes, though further research and development are needed to expand their applications (Arjunan et al., 2023).

6.3.1.2 Wind Energy Harnessing Systems Developments

Wind energy utilizes wind turbines to transform the kinetic energy of the wind into electrical power. Recent advancements in wind turbine technology have significantly contributed to the progression of a low-carbon future. These advancements have ended up in the creation of more powerful and economical turbines that can capture and transform a higher quantity of wind energy into electricity. Rotor designs, blade materials, and tower heights have evolved to increase energy yield and reduce costs. Offshore wind farms, situated in coastal regions or offshore waterways, get advantages from more powerful and constant wind velocities, leading to increased capacity factors and energy production. Floating wind turbines represent a recent innovation in offshore wind energy, enabling deployment in deeper waters and expanding the potential for offshore wind development. Additionally, advancements in wind turbine control systems, such as pitch and yaw control, optimize performance and enhance grid integration. Smart wind farms utilize algorithms for machine learning and data analytics to optimize turbine operation, predict maintenance needs, and mitigate risks (Díaz-González et al., 2012). Overall, wind energy developments continue to drive down costs and increase competitiveness, positioning wind power as a leading renewable energy source (Figure 6.7).

Small wind turbines (SWTs) are a growing alternative to traditional wind farms, ideal for distributed energy needs. They harness wind energy efficiently due to their adaptable size and location. Recent advancements focus on miniaturization and improved energy harvesting, involving three stages: converting wind to mechanical energy, transforming it into electrical energy, and storing or using it directly (Wang et al., 2023). Integrating multiple power generation mechanisms (PGMs) may improve the power density, efficiency, and use of space of SWTs, resulting in more

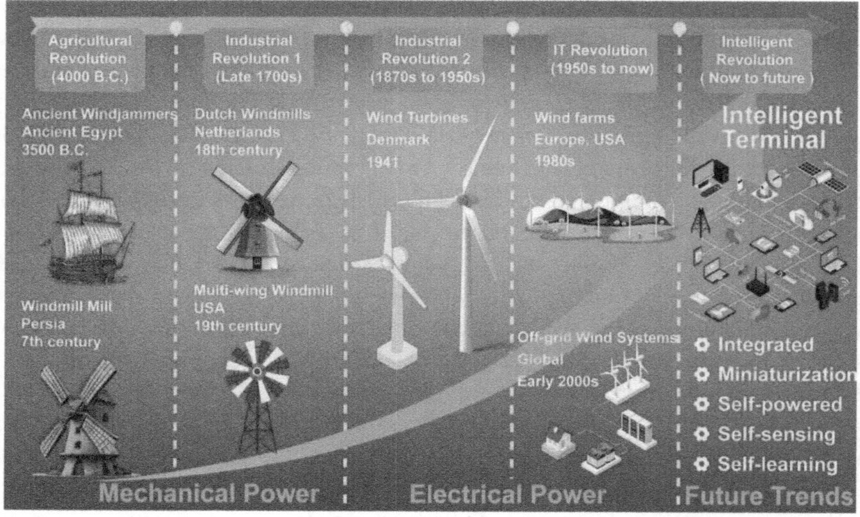

FIGURE 6.7 Key milestones in wind energy technology development (Wang et al., 2023).

stable and reliable performance and a broader operating range. Researchers have developed many hybrid integrated generators (HIGs) for SWTs by combining electromagnetic generators (EMG) with piezoelectric generators (PEG) or triboelectric nanogenerators (TENG), leveraging EMG's high current output to improve overall performance (Wang et al., 2023) (Figure 6.8).

6.3.1.3 Biomass and Bioenergy

Biomass and bioenergy technologies generate heat, power, and transportation fuels from agricultural leftovers, forestry residues, and organic waste. Biochemical or thermochemical methods transform biomass into biofuels. Biochemical processes like fermentation create ethanol and biodiesel from sugars or starches. Gasification and pyrolysis turn biomass into syngas or bio-oil, which may be used to make fuels or generate heat and power. Recent innovations in biomass and bioenergy technologies focus on increasing efficiency, reducing emissions, and expanding feedstock options. Advanced biofuels, derived from nonfood feedstocks such as algae, and cellulosic biomass, offer improved sustainability and lower lifecycle emissions compared to first-generation biofuels (Adams et al., 2017). Integrated biorefineries, which produce a range of bio-based products alongside biofuels, enhance the economic viability of bioenergy production. Additionally, co-firing biomass with coal in power plants reduces carbon emissions and promotes the use of renewable energy (Bhardwaj et al., 2017) (Figure 6.9).

Innovations in bioenergy conversion significantly enhance the synergy between bioenergy and other bio-based products, improving economic and environmental performance. These advancements can be applied through on-site co-production in biorefineries or across multiple locations via interconnected biomass supply networks. For instance, sawmills that process stem wood for construction materials

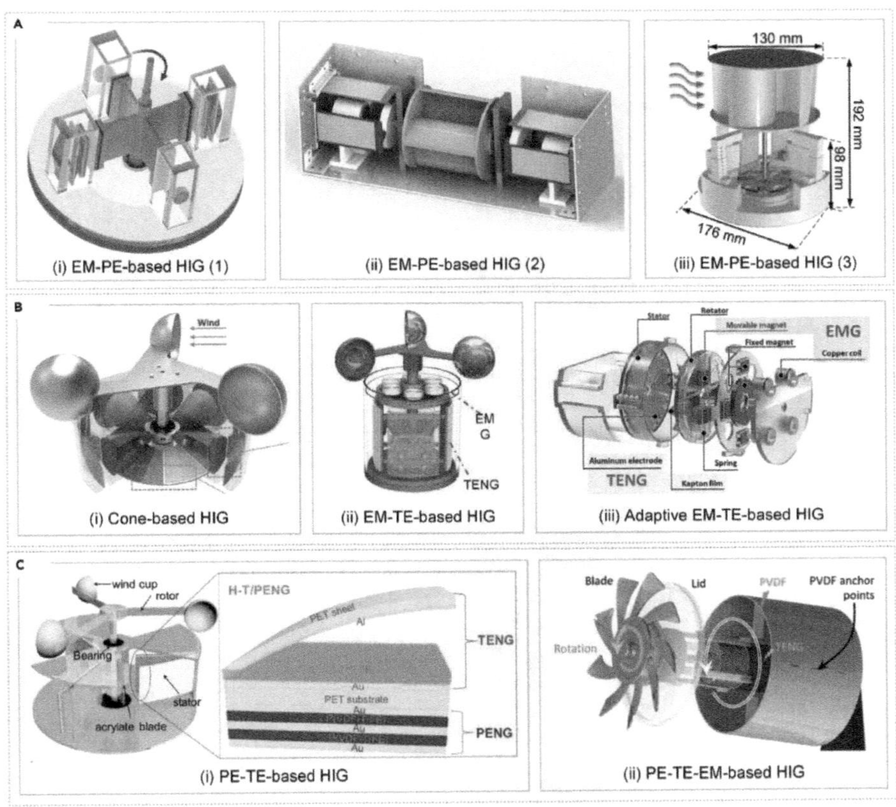

FIGURE 6.8 Different hybrid integrated generators for small wind turbine (Wang et al., 2023).

produce residues such as bark, wood chips, and sawdust. Modern bioenergy technologies convert these residues into energy, enhancing the sawmills' economic and environmental sustainability by reducing waste and providing a renewable energy source (IEA BIOENERGY, 2022).

In the food processing industry, substantial biomass residues are generated from processes such as olive oil production, dairy, and meat processing. Innovations in bioenergy conversion allow these residues to be transformed into energy and valuable biochemicals, improving overall industry efficiency and creating additional revenue streams. Biorefineries exemplify the integration of these innovations, handling diverse biomass inputs to produce energy, biofuels, and biochemicals. Co-locating biorefineries with other industrial operations lead to significant efficiencies by turning by-products into feedstock. Innovations like anaerobic digestion, gasification, and pyrolysis are revolutionizing the sector. Anaerobic digestion converts organic waste into biogas, while gasification and pyrolysis transform solid biomass into syngas or bio-oil, which can be transformed further into high-value fuels and chemicals (Tshikovhi & Motaung, 2023).

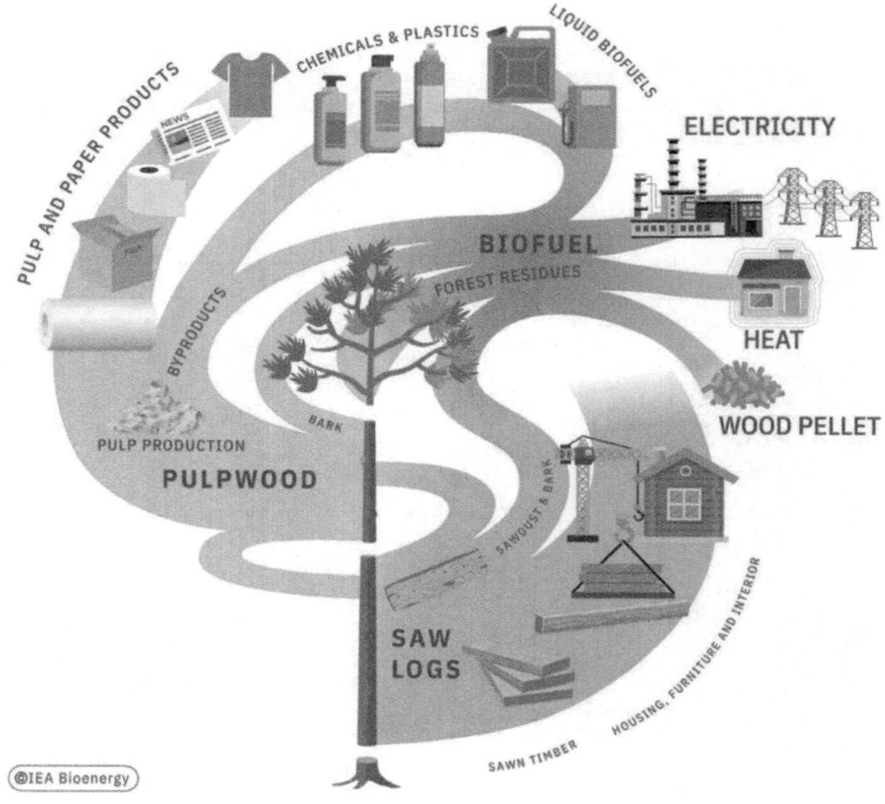

FIGURE 6.9 Utilization of wood residues in a bioeconomy (IEA Bioenergy, 2022).

6.3.2 Energy Storage Technologies

6.3.2.1 Battery Technologies

Battery technologies play a pivotal role in storing electricity generated from renewable energy sources and facilitating its use when needed. Among various types, LIBs stand out as the most prevalent rechargeable batteries used in diverse applications, including EVs, grid storage systems, and portable electronics. Recent strides in LIB technology have been geared toward enhancing energy density, extending cycle life, and fortifying safety while concurrently driving down costs. Moreover, exploration into alternative battery chemistries, such as sodium-ion, zinc-air, and solid-state batteries, is underway to address scalability, sustainability, and performance concerns inherent in LIBs (Manthiram, 2017).

The global battery market is expected to expand quickly, expected to surpass 2,500 GWh in the next decade. Electric mobility is a key driver, influencing battery demand across various applications and regions (Figure 6.10). China, a major player in battery manufacturing, is anticipated to decrease its global production share, while the rest of the world (RoW) will likely see an increase. Government policies promoting electric and zero-emission vehicles are spurring advancements in

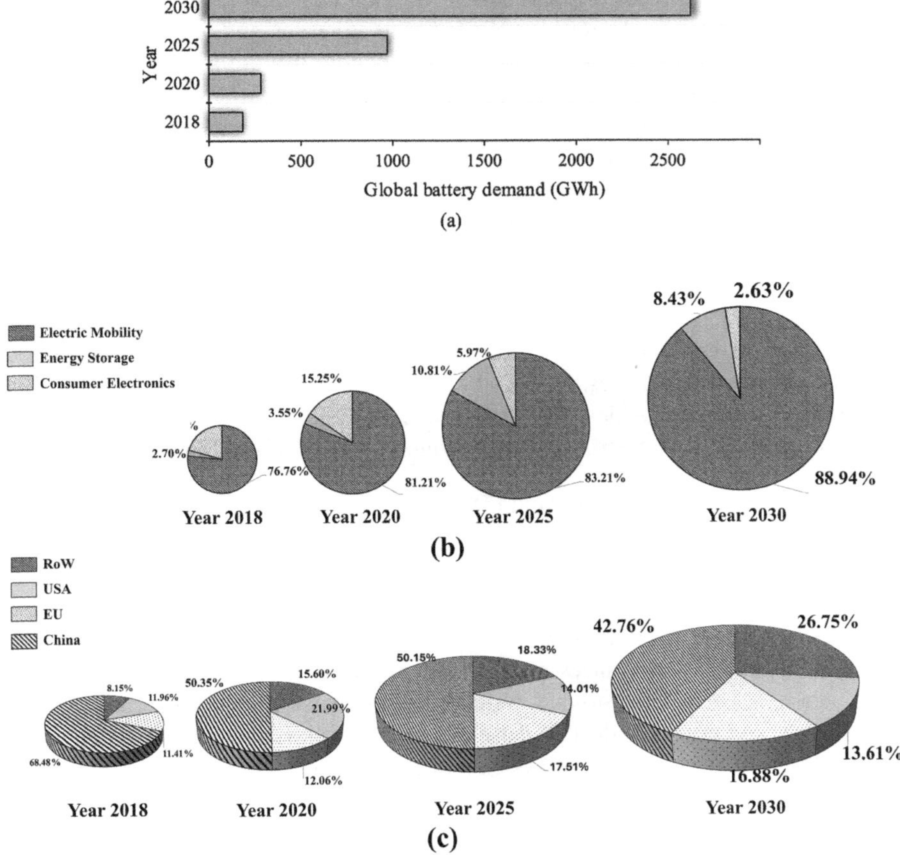

FIGURE 6.10 Global battery market (a) development. (b) Application-specific demands. (c) Regional demands (Liu et al., 2022).

battery materials and smart mobility technologies. China aims to peak emissions by 2030, with the USA targeting 50% of new vehicles to be emission free by 2030, and Europe aiming for nearly all vehicles to achieve zero emissions by 2035. Achieving an energy density of 500 Wh/kg in LIBs for EV applications is still an important constraint for contemporary battery technology (Liu et al., 2022).

Nickel-based technologies, such as nickel-zinc (Ni-Zn) and nickel metal hydride (Ni-MH) batteries, have propelled the battery industry into a new phase. Ni-Zn batteries possess a notable capacity for storing energy per unit mass and are economically advantageous due to their cheap material expenses. However, they are hindered by a limited number of charge-discharge cycles, which restricts their practicality for commercial use. On the other hand, Ni-MH batteries, which have a specific energy of up to 80 Wh/kg and an energy density of 250 Wh/L, have been used in both BEVs and HEVs. Prior to the introduction of lithium-based batteries, lead-acid and nickel-based batteries were the predominant types. Lithium-metal and LIBs have become widely used, but their specific energy and density have certain limits. As a result, researchers have been exploring alternative technologies such as metal/air, sodium-beta, sodium-ion, zinc-ion, and magnesium-ion batteries. Notable advancements in technology include lithium/air, lithium/sulphur, and solid-state batteries. Additionally, dual-ion and dual-carbon batteries have the potential for use in stationary energy storage. Advanced management technologies include several battery models such as electrochemical, equivalent circuit, integral-order, fractional-order, and data-driven models. These technologies also incorporate safety systems like heat management, cell balancing, and fault diagnostics. In the future, LIBs will continue to be the dominant choice for EV applications. However, lithium metal is considered the most promising material for the anode in future improvements. This is despite the problems posed by the complicated lithium-oxygen system, which necessitates substantial study. Alternative ion batteries, namely sodium-ion batteries, are also becoming increasingly popular and have the potential to work alongside LIBs in future EV applications. This emphasizes the wide range of battery chemistries and technologies available (Liu et al., 2022).

6.3.2.2 Thermal Energy Storage (TES)

TES systems are designed to store heat derived from renewable energy sources or surplus heat from industrial processes, releasing it as needed for heating, cooling, or power generation purposes. TES technologies consist of many ways, such as sensible heat storage, latent heat storage, and thermochemical storage. Sensible heat storage systems employ substances such as water or molten salts to collect and retain heat, whereas latent heat storage systems utilize phase-change materials (PCMs) that absorb or release heat during phase transitions (Kalaiselvam & Parameshwaran, 2014). Thermochemical storage systems operate via reversible chemical reactions to store and release heat, offering high energy density and extended storage durations. TES solutions significantly bolster the efficiency and flexibility of renewable energy systems by enabling dispatchable power generation and facilitating load shifting (Dincer & Rosen, 2011). TES is essential for integrating sectors and optimizing energy use in a scenario aiming to keep the rise in global temperature to less than 2°C. As the electrification of heating, cooling, and transport sectors grows, TES helps manage increased electricity demand (projected to reach 49% of final energy by 2050). It balances the grid by storing excess electricity during low-demand periods for later use, reducing strain on power generation and enhancing overall system efficiency as indicated in Figure 6.11. TES also supports renewable energy integration by mitigating intermittency issues, ensuring a reliable energy supply. TES is pivotal in converting to a low-carbon future by optimizing energy use and assisting renewable energy integration (IRENA, 2020a).

Figure 6.12 indicates TES technologies their operating temperatures and durations. High-temperature solutions like chemical looping (800°C–1,000°C) and CPCMs (300°C–700°C) offer long-term storage from days to months. Molten salts (150°C–600°C) and solid-state (200°C–400°C) systems provide versatile short- to long-term storage. Medium-term options include salt hydration (30°C–200°C) and absorption systems (50°C–150°C), ideal for daily to seasonal storage. Low-temperature PCMs (–30°C to 150°C), WTTES (5°C–95°C), and UTES (5°C–95°C) are efficient for short- to medium-term applications. Ice storage (0°C) and sensible

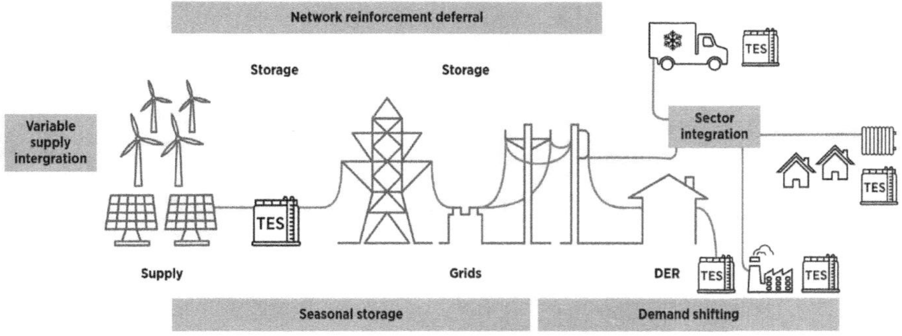

FIGURE 6.11 Major applications of TES in the energy sector (IRENA, 2020a).

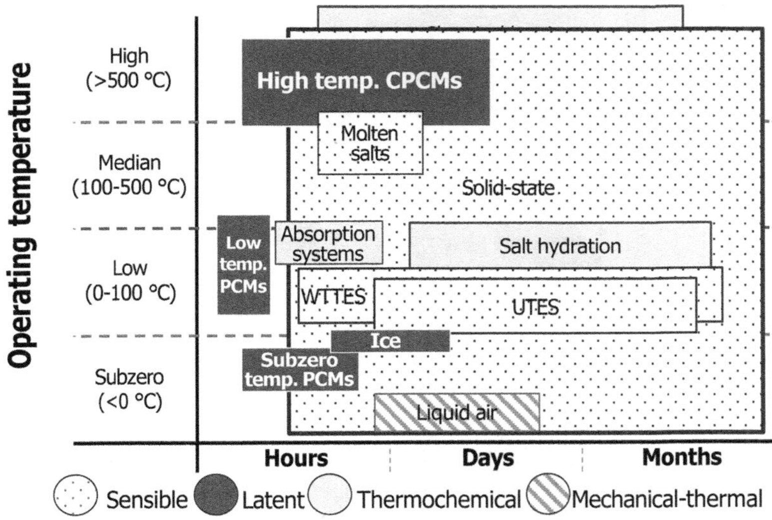

FIGURE 6.12 Working temperatures and duration for TES technologies (IRENA, 2020a). cPCM, composite phase-change material; PCM, phase-change material; WTTS, water tank thermal energy storage.

heat storage (up to 400°C) are suitable for short-term needs, while liquid air storage (–196°C to –150°C) covers short- to long-term demands. These technologies enable flexible and sustainable energy solutions across various sectors (IRENA, 2020a).

6.3.2.3 Hydrogen Storage and Fuel Cells

Hydrogen and fuel cells are clean, adaptable energy storage solutions for transportation, power generation, and industry. Compressed gas, liquid hydrogen, and solid-state compounds like metal hydrides and carbon can store hydrogen. The most common element, hydrogen, accounts for 90% of atoms and 75% of mass in normal matter. It transports pure energy. Fuel cell systems electrochemically oxidize hydrogen gas to produce clean water and no carbon dioxide. Hydrogen has become an energy vector for processing

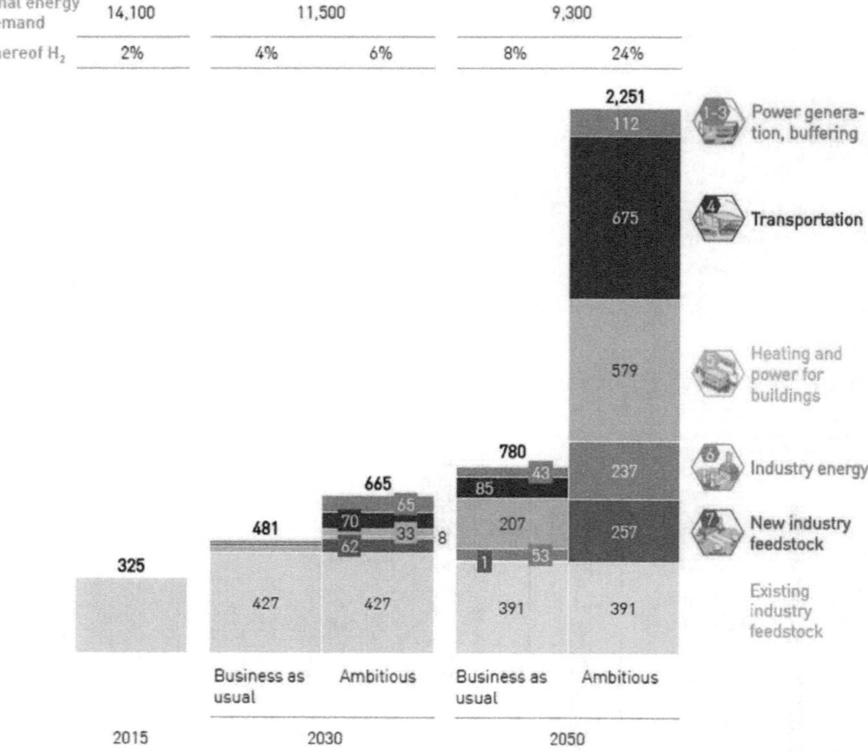

FIGURE 6.13 Projected energy demand and hydrogen deployment in Europe (2015–2050) (Fan et al., 2021).

ammonia, methanol, and petroleum in addition to its industrial feedstock role. Hydrogen is in demand in transportation, power generation, and militarized equipment because of its effectiveness and low emissions. Numerous countries have created hydrogen roadmaps as a result of the rapid advancement of hydrogen technology and the increase in energy demand. Fuel cells and hydrogen can provide many of the energy operations and fulfil growing societal requirements. Thus, several governments consider the development of hydrogen in their goals and promote the fuel cell sector (Fan et al., 2021). Figure 6.13 indicates the roadmap planned for Europe from 2015 to 2050.

Hydrogen with oxygen electrochemically reacts to generate energy, producing only water vapour. PEMFCs and SOFCs are prominent types used for stationary and mobile applications. Recent advancements centre on enhancing fuel cell performance, durability, and cost-effectiveness, driving commercialization and deployment (Felseghi et al., 2019; Manoharan et al., 2019; Singla et al., 2021). Clean energy sources, notably hydrogen energy, are expected to become more popular in the future decades to reduce environmental deterioration and fossil fuel dependence. Hydrogen and electrification will help decarbonize global energy systems. Liquid hydrogen carriers reduce infrastructure needs due to their increased volumetric density and simplicity of storage and movement. Reviewing relevant literature yields three fuel cell design, commercial market, and strategic support (Fan et al., 2021).

6.3.2.3.1 *Issues with Commercial Release*

Stability and cost are major barriers to hydrogen fuel cell commercialization. Overcoming these issues requires technical support for large-scale, cost-effective, dependable applications. Fuel cells must be more efficient, and hydrogen generation and storage must improve for safety and stability. Durable and efficient hydrogen generation, storage, transit, and use should be integrated.

6.3.2.3.2 *Fuel Cell Vehicle Market Potential*

Fuel cell vehicles (FCVs) are commercially desirable due to their zero emissions. For buses and big trucks, FCVs outperform EVs in range and refuelling time. FCVs are promising for city, railway, seaport, and warehouse transportation. FCV systems need to reduce prices, improve stability and performance, and construct more hydrogen refuelling stations (HRSs) to gain market share.

6.3.2.3.3 *Policy and Strategic Support*

Strong, concentrated, and long-term energy strategies and government policies are needed to embrace hydrogen technologies. These efforts must disrupt energy systems and advance low-carbon technology. Regional coordinating organizations should outline hydrogen policy and correct country imbalances. A comprehensive policy system is needed to promote the evolution of hydrogen and fuel cell technologies, ensuring they achieve performance and cost requirements while providing stability and vitality. These advances and strategic focuses will shape hydrogen energy and its position in a sustainable, decarbonized global energy system.

Hydrogen is mostly employed in oil, ammonia, and metal processing and as a clean energy source for internal combustion engines, turbines, and FCVs (as indicated in Figure 6.14). Hydrogen storage methods convert surplus renewable energy

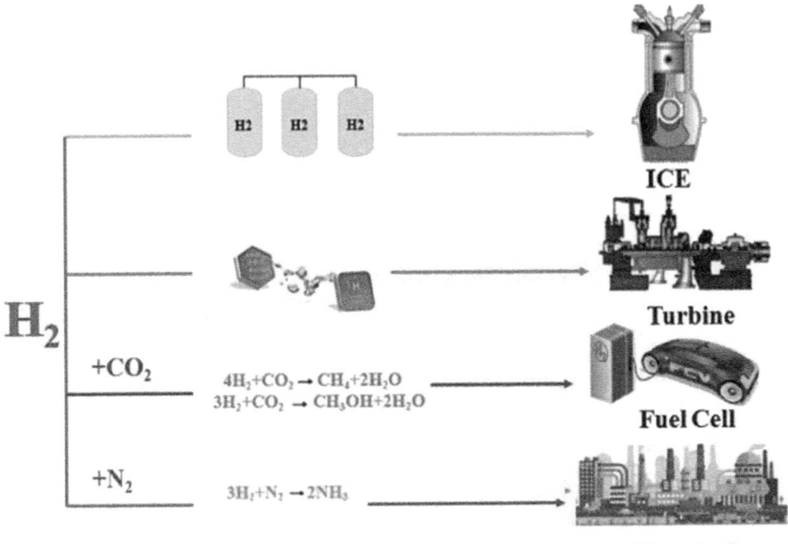

FIGURE 6.14 Hydrogen storage and utilization pathways (Fan et al., 2021).

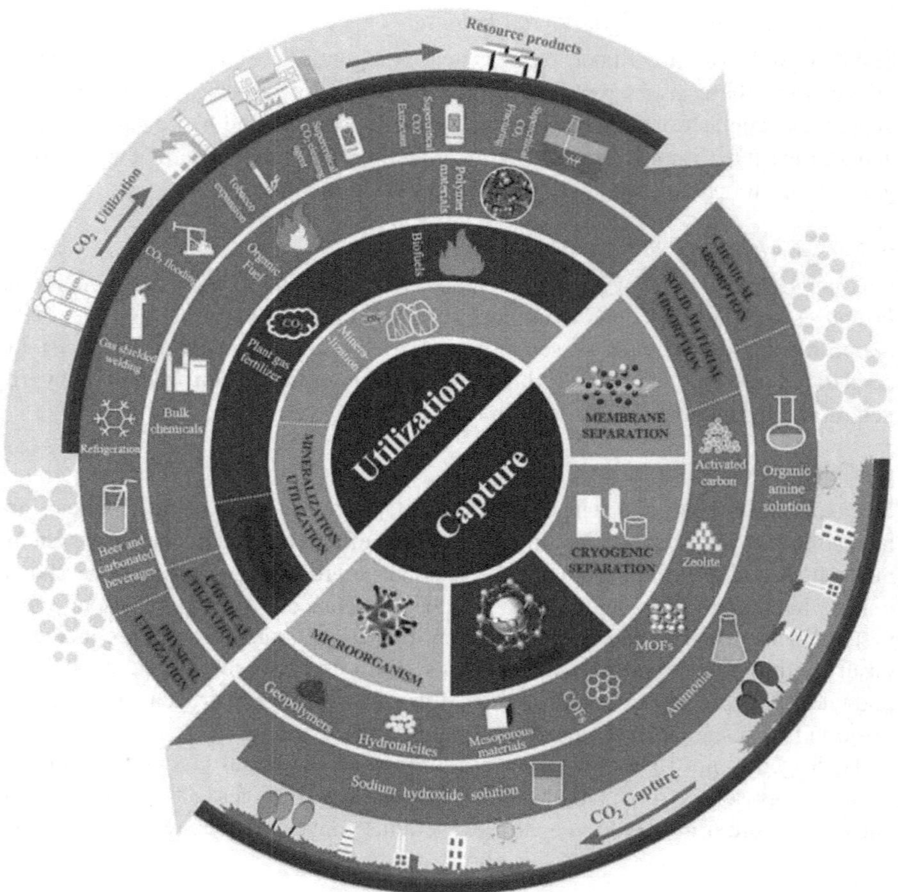

FIGURE 6.15 Schematic diagram of CO_2 capture and utilization cycle (Fu et al., 2022).

like solar or wind power into hydrogen via electrolysis. A flexible energy carrier, hydrogen may be stored and used across industries. The Power-to-Gas (P2G) process converts surplus power into hydrogen, which may be pumped into gas networks, stored, or utilized in fuel cells to generate electricity during peak demand. Fuel cell cars use clean HRSs, which release only water vapour. Hydrogen is used in fossil fuel-dependent industries like ammonia and petroleum refining. Hydrogen can be stored in tanks or pumped into natural gas systems to mix with methane or distribute energy in specialized hydrogen grids. These approaches improve renewable energy efficiency and dependability, reduce carbon emissions, and promote sustainable energy (Staffell et al., 2019).

6.3.3 Carbon Capture and Utilization

6.3.3.1 Carbon Capture Techniques

The increasing levels of CO_2 in the atmosphere due to industrial activities cause a significant threat to the environment and lead to climate change. To mitigate these effects, effective carbon capture and utilization methods are to be adopted. Figure 6.15 outlines various CO2 capture technologies, including absorption, adsorption, membrane separation, cryogenic separation, hydrate technology and microbial technology. For utilization, different approaches are indicated in Figure 6.15, such as physical, chemical, biological, and mineralization. Carbon capture methods gather CO_2 from factories, power plants, and other stationary sources before releasing them. Primary techniques include postcombustion capture, precombustion capture, and oxy-fuel combustion (Boot-Handford et al., 2014) shown in Figure 6.16. Postcombustion capture removes CO_2 from flue gases using chemical solvents or adsorbents. Precombustion capture segregates CO_2 from syngas produced by fossil fuel or biomass gasification. Oxy-fuel combustion burns fuels in an oxygen-rich atmosphere, facilitating CO_2 capture. Recent innovations focus on enhancing capture efficiency, reducing energy requirements, and lowering costs to enhance commercial viability (Boot-Handford et al., 2014). Precombustion capture requires less regenerative energy and smaller equipment but suffers from high costs and operational limitations. Oxy-fuel combustion capture offers benefits like high CO_2 concentration and reduced NO_x emissions but demands significant oxygen, increasing costs and energy use. Postcombustion capture is notable for its lower initial costs, rapid deployment, and operational flexibility,

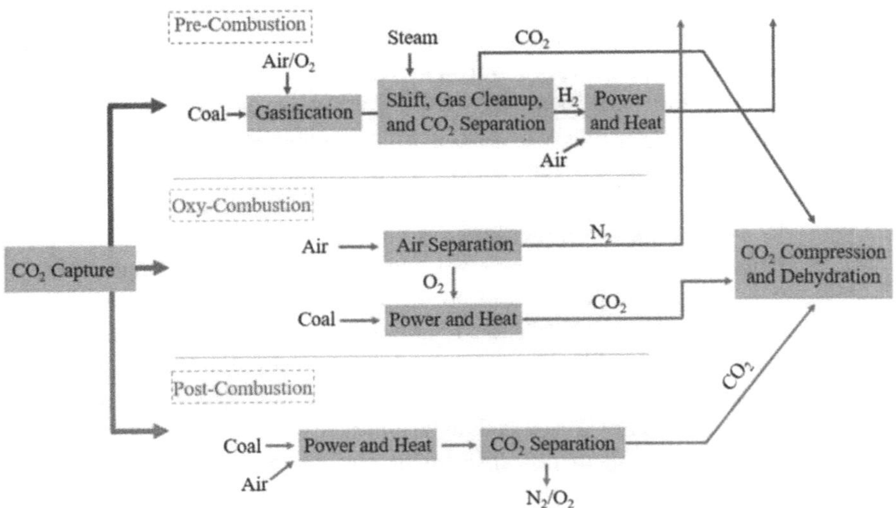

FIGURE 6.16 Carbon capture technologies for power plants (Fu et al., 2022).

making it well-established in the market. However, it faces challenges of higher energy consumption due to lower CO_2 partial pressure and requires large-scale equipment. These methods highlight distinct trade-offs between efficiency, cost, and scalability, underscoring ongoing research needs to optimize these technologies for broader adoption in combatting climate change.

6.3.3.2 Carbon Utilization Methods

Carbon utilization methods to transform collected CO_2 into useful goods, lowering GHG emissions and promoting a circular carbon economy. Direct air capture techniques capture CO_2 directly from ambient air for utilization or storage (Atanu Mukherjee & Chatterjee, 2022). CO_2 can be changed into fuels such as methane, methanol, or synthetic hydrocarbons through chemical processes like hydrogenation or electrochemical reduction. Additionally, CO_2 mineralization into stable carbonates or use as a feedstock for carbon-based materials like graphene offers opportunities for emissions mitigation and value-added product creation.

Carbon utilization methods encompass both direct and indirect techniques to mitigate CO_2 emissions and create valuable products. Here's a concise summary of these methods (Yusuf & Ibrahim, 2023):

1. **Direct Utilization Techniques**
   - **Iron and Steel Making:** CO_2 regulates molten steel temperature, inhibits oxidation, and controls dust during steel manufacturing, including iron ore reduction and coke combustion.
   - **Alumina Industry:** In the Bayer process, CO_2 neutralizes the caustic nature of red mud, a toxic byproduct of bauxite digestion, offering a low-cost and highly available solution.
   - **Rare Earths Recovery from Acid Mine Drainage:** CO_2 helps recover valuable rare earth elements derived from acid mine effluent, a byproduct of the oxygen-mediated weathering of sulphide minerals.
   - **Enhanced Oil Recovery (EOR):** High-pressure CO_2 injection reduces oil viscosity and interfacial tension, promoting volumetric swelling and facilitating higher oil recovery in reservoirs.
   - **Concrete Building Materials:** Concrete's carbonation curing process, which uses CO_2, produces carbonate materials (such $CaCO_3$ or $MgCO_3$) that affect the longevity and strength of things made of concrete.
2. **Indirect Utilization Techniques**
   - **Thermochemical Conversion of CO_2:** In this process, a enough amount of energy is provided in the presence of a substance that reduces the CO_2, resulting in the transformation of CO_2 into fuels and chemicals.
     - **Dry Reforming of Methane:** Converts CO_2 and methane into synthesis gas (a mixture of H_2 and CO), which can be utilized to make different chemicals and fuels.
     - **Oxy-CO_2 Reforming of Methane:** Uses oxygen and CO_2 to convert methane into synthesis gas, with the added benefit of improved efficiency and reduced energy consumption.

- **Electrochemical Reduction of CO_2:** Uses electrical energy to convert CO_2 into useful chemicals such as carbon monoxide, formic acid, and hydrocarbons, offering a sustainable way to recycle CO_2.
- **Photocatalytic Reduction of CO_2:** Utilizes light energy to convert CO_2 into fuels and chemicals through photocatalysts, providing a renewable energy-driven solution for CO_2 conversion.
- **Biological Conversion of CO_2:** Employs microorganisms such as algae and bacteria to convert CO_2 into organic compounds, including biofuels and bioplastics, through photosynthesis or other metabolic pathways.

6.3.3.2.1 Storage Solutions

The significance of CO_2 storage has increased significantly after the Paris Agreement was signed in 2015, leading to a rush in research focused on finding viable storage methods. It may be roughly classified into natural and man-made methods. Natural approaches are represented by terrestrial sequestration, whereas man-made tactics are encompassed by geological formations. Important approaches include terrestrial and marine sequestration, capturing CO_2 through gas hydrates, and geological processes such as storing it in saline aquifers, abandoned oil and gas reservoirs, coal seams, and during EOR. Every method has distinct advantages and difficulties, emphasizing continuous endeavours to enhance these technologies for efficient carbon management and ecological sustainability (Yusuf & Ibrahim, 2023).

6.4 CONCLUSION

This book chapter emphasizes the necessity of continuous eco-innovation and proactive environmental protection for sustainable development. Societies can promote economic growth and employment creation while mitigating climate risks by prioritizing low-carbon technologies and adopting renewable energy solutions. The chapter emphasizes the importance of implementing targeted measures and comprehensive policy frameworks in key sectors to accelerate the worldwide shift towards a sustainable future. It suggests the transformative potential of technological advancements in renewable energy, highlighting their important role in the achievement of net-zero emissions by 2050. Ultimately, stakeholders can ensure a resilient, prosperous future for future generations by effectively addressing climate challenges through collaborative efforts and integrated approaches.

REFERENCES

Adams, P., Bridgwater, T., Lea-Langton, A., Ross, A., & Watson, I. (2017). Biomass conversion technologies. In *Greenhouse Gas Balances of Bioenergy Systems*. Elsevier, Amsterdam, The Netherlands. https://doi.org/10.1016/B978-0-08-101036-5.00008-2

Albino, V., Ardito, L., Dangelico, R. M., & Messeni Petruzzelli, A. (2014). Understanding the development trends of low-carbon energy technologies: a patent analysis. *Applied Energy*, *135*, 836–854. https://doi.org/10.1016/j.apenergy.2014.08.012

Arjunan, T. V., Selvaraj, V., & Matheswaran, M. M. (2023). *Solar Thermal Conversion Technologies for Industrial Process Heating*. CRC Press, Boca Raton, FL. https://doi.org/10.1201/9781003263326

Bhardwaj, A. K., Chauhan, R., Kumar, R., Sethi, M., & Rana, A. (2017). Experimental investigation of an indirect solar dryer integrated with phase change material for drying *Valeriana jatamansi* (medicinal herb). *Case Studies in Thermal Engineering*, *10*, 302–314. https://doi.org/10.1016/j.csite.2017.07.009

Boot-Handford, M. E., Abanades, J. C., Anthony, E. J., Blunt, M. J., Brandani, S., Mac Dowell, N., Fernández, J. R., Ferrari, M. C., Gross, R., Hallett, J. P., Haszeldine, R. S., Heptonstall, P., Lyngfelt, A., Makuch, Z., Mangano, E., Porter, R. T. J., Pourkashanian, M., Rochelle, G. T., Shah, N., & Fennell, P. S. (2014). Carbon capture and storage update. *Energy and Environmental Science*, *7*(1), 130–189. https://doi.org/10.1039/c3ee42350f

Creutzig, F., Jochem, P., Edelenbosch, O. Y., Mattauch, L., Van Vuuren, D. P., McCollum, D., & Minx, J. (2015). Transport: a roadblock to climate change mitigation? *Science*, *350*(6263), 911–912. https://doi.org/10.1126/science.aac8033

Dada, M., & Popoola, P. (2023). Recent advances in solar photovoltaic materials and systems for energy storage applications: a review. *Beni-Suef University Journal of Basic and Applied Sciences*, *12*(1), 5. https://doi.org/10.1186/s43088-023-00405-5

Díaz-González, F., Sumper, A., Gomis-Bellmunt, O., & Villafáfila-Robles, R. (2012). A review of energy storage technologies for wind power applications. *Renewable and Sustainable Energy Reviews*, *16*(4), 2154–2171. https://doi.org/10.1016/j.rser.2012.01.029

Dietz, T., Gardner, G. T., Gilligan, J., Stern, P. C., & Vandenbergh, M. P. (2009). Household actions can provide a behavioral wedge to rapidly reduce US carbon emissions. *Proceedings of the National Academy of Sciences of the United States of America*, *106*(44), 18452–18456. https://doi.org/10.1073/pnas.0908738106

Dincer, I., & Rosen, M. A. (2011). *Thermal Energy Storage: Systems and Applications*. John Wiley & Sons, New York.

Fan, L., Tu, Z., & Chan, S. H. (2021). Recent development of hydrogen and fuel cell technologies: a review. *Energy Reports*, *7*, 8421–8446. https://doi.org/10.1016/j.egyr.2021.08.003

Fang, X., Misra, S., Xue, G., & Yang, D. (2012). Smart grid - The new and improved power grid: a survey. *IEEE Communications Surveys and Tutorials*, *14*(4), 944–980. https://doi.org/10.1109/SURV.2011.101911.00087

Felseghi, R. A., Carcadea, E., Raboaca, M. S., Trufin, C. N., & Filote, C. (2019). Hydrogen fuel cell technology for the sustainable future of stationary applications. *Energies*, *12*(23), 12234593. https://doi.org/10.3390/en12234593

Fu, L., Ren, Z., Si, W., Ma, Q., Huang, W., Liao, K., Huang, Z., Wang, Y., Li, J., & Xu, P. (2022). Research progress on CO_2 capture and utilization technology. *Journal of CO_2 Utilization*, *66*(July), 102260. https://doi.org/10.1016/j.jcou.2022.102260

Gamil, A., Li, P., Ali, B., & Hamid, M. A. (2022). Concentrating solar thermal power generation in Sudan: potential and challenges. *Renewable and Sustainable Energy Reviews*, *161*(February), 112366. https://doi.org/10.1016/j.rser.2022.112366

Griscom, B. W., Adams, J., Ellis, P. W., Houghton, R. A., Lomax, G., Miteva, D. A., Schlesinger, W. H., Shoch, D., Siikamäki, J. V., Smith, P., Woodbury, P., Zganjar, C., Blackman, A., Campari, J., Conant, R. T., Delgado, C., Elias, P., Gopalakrishna, T., Hamsik, M. R., & Fargione, J. (2017). Natural climate solutions. *Proceedings of the National Academy of Sciences of the United States of America*, *114*(44), 11645–11650. https://doi.org/10.1073/pnas.1710465114

IEA. (2020). *Energy Technology Perspectives 2020*. International Energy Agency, Paris, France. https://doi.org/10.1787/ab43a9a5-en

IEA BIOENERGY. (2022). How Bioenergy Contributes to a Sustainable Future. In *IEA Bioenergy Report 2023*. https://www.ieabioenergyreview.org/wp-content/uploads/2022/12/IEA_BIOENERGY_REPORT.pdf

International Energy Agency. (2021). Net Zero by 2050: A Roadmap for the Global Energy Sector. 70. International Energy Agency, Paris, France.

IPCC. (2018). *Global Warming of 1.5°C.* In *Special Report on Global Warming of 1.5°C.* Cambridge University Press, Cambridge, MA.

IPCC. (2022). *Climate Change 2022- Mitigation of Climate Change - Full Report.* Cambridge University Press, Cambridge, MA.

IRENA. (2019). *Global Energy Transformation: A Roadmap to 2050.* International Renewable Energy Agency, Abu Dhabi.

IRENA. (2020a). *Innovation Outlook: Thermal Energy Storage.* International Renewable Energy Agency, Abu Dhabi. https://www.irena.org/publications/2020/Nov/Innovation-outlook-Thermal-energy-storage

IRENA. (2020b). *Renewable Power Generation Costs in 2019.* International Renewable Energy Agency, Abu Dhabi. https://www.irena.org/-/media/Files/IRENA/Agency/Publication/2018/Jan/IRENA_2017_Power_Costs_2018.pdf

Kalaiselvam, S., & Parameshwaran, R. (2014). *Thermal Energy Storage Technologies for Sustainability: Systems Design, Assessment and Applications.* Elsevier, Amsterdam, The Netherlands.

Liu, W., Placke, T., & Chau, K. T. (2022). Overview of batteries and battery management for electric vehicles. *Energy Reports, 8*, 4058–4084. https://doi.org/10.1016/j.egyr.2022.03.016

Liu, Z., Deng, Z., Davis, S., & Ciais, P. (2023). Monitoring global carbon emissions in 2022. *Nature Reviews Earth and Environment, 4*(4), 205–206. https://doi.org/10.1038/s43017-023-00406-z

Liu, Z., Deng, Z., Davis, S. J., Giron, C., & Ciais, P. (2022). Monitoring global carbon emissions in 2021. *Nature Reviews Earth and Environment, 3*(4), 217–219. https://doi.org/10.1038/s43017-022-00285-w

Manoharan, Y., Hosseini, S. E., Butler, B., & Alzhahrani, H. (2019). Hydrogen fuel cell vehicles: current status and future prospect. *Applied Science, 9*(11), 2296. https://doi.org/https://doi.org/10.3390/app9112296

Manthiram, A. (2017). An outlook on lithium ion battery technology. *ACS Central Science, 3*(10), 1063–1069. https://doi.org/10.1021/acscentsci.7b00288

Maréchal, K. (2007). The economics of climate change and the change of climate in economics. *Energy Policy, 35*(10), 5181–5194. https://doi.org/10.1016/j.enpol.2007.05.009

Mukherjee, A., & Chatterjee, S. (2022). Carbon capture utilization and storage strategy, policy framework and its deployment mechanism in India. *Niti Aayog, November*, 168. https://www.niti.gov.in/sites/default/files/2022-12/CCUS-Report.pdf%0Awww.niti.gov.in

Olabi, A. G. (2017). Renewable energy and energy storage systems. *Energy, 136*, 1–6. https://doi.org/10.1016/j.energy.2017.07.054

Pérez-Lombard, L., Ortiz, J., & Pout, C. (2008). A review on buildings energy consumption information. *Energy and Buildings, 40*(3), 394–398. https://doi.org/10.1016/j.enbuild.2007.03.007

Sanz-Pérez, E. S., Murdock, C. R., Didas, S. A., & Jones, C. W. (2016). Direct capture of CO_2 from ambient air. *Chemical Reviews, 116*(19), 11840–11876. https://doi.org/10.1021/acs.chemrev.6b00173

Shubbak, M. H. (2019). Advances in solar photovoltaics: technology review and patent trends. *Renewable and Sustainable Energy Reviews, 115*(July), 109383. https://doi.org/10.1016/j.rser.2019.109383

Singh, G., Abdul, A. P. J., & Singh, B. R. (2019). Recent developments in hybrid solar power generation: a review. *International Journal of Engineering Research & Technology, 8*(6), 1019–1031. www.ijert.org

Singla, M. K., Nijhawan, P., & Oberoi, A. S. (2021). Hydrogen fuel and fuel cell technology for cleaner future: a review. *Environmental Science and Pollution Research, 28*(13), 15607–15626. https://doi.org/10.1007/s11356-020-12231-8

Staffell, I., Scamman, D., Velazquez Abad, A., Balcombe, P., Dodds, P. E., Ekins, P., Shah, N., & Ward, K. R. (2019). The role of hydrogen and fuel cells in the global energy system. *Energy and Environmental Science*, *12*(2), 463–491. https://doi.org/10.1039/c8ee01157e

Stavins, R. N. (2008). A meaningful US cap-and-trade system to address climate change. *Harvard Environmental Law Review*, *32*, 293.

Tshikovhi, A., & Motaung, T. E. (2023). Technologies and innovations for biomass energy production. *Sustainability (Switzerland)*, *15*(16), 1–21. https://doi.org/10.3390/su151612121

Vodapally, S. N., & Ali, M. H. (2022). A comprehensive review of solar photovoltaic (PV) technologies, architecture, and its applications to improved efficiency. *Energies*, *16*(1), 319. https://doi.org/10.3390/en16010319

Wang, H., Xiong, B., Zhang, Z., Zhang, H., & Azam, A. (2023). Small wind turbines and their potential for internet of things applications. *IScience*, *26*(9), 107674. https://doi.org/10.1016/j.isci.2023.107674

World Health Organization. (2021). *WHO Global Air Quality Guidelines: Particulate Matter (PM2.5 and PM10), Ozone, Nitrogen Dioxide, Sulfur Dioxide and Carbon Monoxide*. World Health Organization, Geneva, Switzerland.

Yusuf, M., & Ibrahim, H. (2023). A comprehensive review on recent trends in carbon capture, utilization, and storage techniques. *Journal of Environmental Chemical Engineering*, *11*(6), 111393. https://doi.org/10.1016/j.jece.2023.111393

7 A Preliminary Assessment of PM1.0 and PM2.5 levels in Madurai, India

Sivaramasundaram Krishnamoorthy, Nagaprasad Nagaraj, Shanmugam Ramaswamy, L. Priyanka Dwarampudi, Krishnaraj Ramaswamy, and S. Venkatesh

7.1 INTRODUCTION

Worldwide comprehensive studies on particulate matter (PM) reveal the strong relationship between PM exposure and increased cardiovascular morbidity [1] and mortality [2]. Past studies reported an increase of 1.22%–13% in the mortality rate for every 10 $\mu g/m^{-3}$ increase in short- and long-terms PM exposure [3]. World Health Organization (WHO) reports suggest that about 30% of respiratory diseases are related to high-level PM exposure [4]. Presently, India is one of the fastest-growing economies in the world and has experienced unprecedented rapid urbanization in recent years. The associated developmental activities, in turn, result in the increased consumption of the energy resources and hence in environmental pollution, of which air is the worst affected medium, dominated by particulate pollution. As per the World Bank data, three of the world's 20 cities having the highest PM_{10} levels were in India, while then Indian cities had the highest $PM_{2.5}$ levels across the globe. Since 1991, it has been clear that many Indian cities fall short of the required levels of air quality and that practically all of them suffer from significant particle pollution. A recent study in Delhi has reported that around 90% of the particulate levels have been accounted for by vehicular, road dust, and construction emissions [5].

$PM_{1.0}$ and $PM_{2.5}$ reported higher impacts against PM_{10} on respiratory health in the elderly, including non-allergic subjects [6]. Using machine learning (ML) approaches such as ordinary least squares land-use regression (OLS-LUR) and random forest land-use regression (RF-LUR) to predict $PM_{1.0}$ with the higher accuracy played a vital role for the area of authority to plan the decision accordingly [7]. A team of researchers applied a positive matrix factorization model to six locations in China for characterizing and spatial source apportionments of ambient PM_{10} and $PM_{2.5}$ [8]. PM_{10} and $PM_{2.5}$ have been uplifted by the presence of SO_x, NO_x, and ammonium, with annual mean 35.49 and 66.58 $\mu g/m^{-3}$ for PM_{10} and $PM_{2.5}$ in urban Beijing [9].

DOI: 10.1201/9781003452072-7

Like many other cities in India, the atmosphere of Madurai city is deteriorating due to rapid urbanization, increased industrialization, transport, population changes, and a rise in production and investments. Though for many years, continuous attention has been given to the assessment of ambient PM_{10} levels, the $PM_{1.0}$ and $PM_{2.5}$ levels have been left unnoticed. So, for the first time, an attempt has been made to assess the ambient $PM_{1.0}$ and $PM_{2.5}$ levels in different parts of the city and the study is dedicated to generating baseline data on these pollutants.

The part of the manuscript is followed by the method section including sample assortment, data analysis, multilinear regression (MLR), and Pearson's correlation (PC) at the level of fundamental illustrations. The next part is followed by results and discussion describing the analysis of hourly variations in the particulate levels, influence of meteorological parameters on PM level, correlation of $PM_{1.0}$ with $PM_{2.5}$, and standard exceedances. This section was followed by summarization of the manuscript in the form of conclusion.

7.2 METHODS

7.2.1 Sample Assortment

Three different geolocations (Goripalayam, Periyar Bus Stand, and Kochadai) have been selected beside the Vaigai River at Madurai, India, for the study (see Figure 7.1). Two stations (Periyar Bus Stand and Kochadai) are south of the river, while one station (Goripalayam) is north of the river. The Vaigai River has episodic extreme climate-related history [10,11]. Nearly 15 km of perimeter have been covered within central Madurai.

Eight-hour sampling (09:00–17:00), including the morning peak traffic hours, was carried out using a commercially available low-cost handheld particulate monitor [12]. The samples were assorted on a weekly basis with 15-minute time resolution at each sampling site from 8th January to 15th February 2023. The hourly averages of the particulate concentrations were calculated from the obtained data.

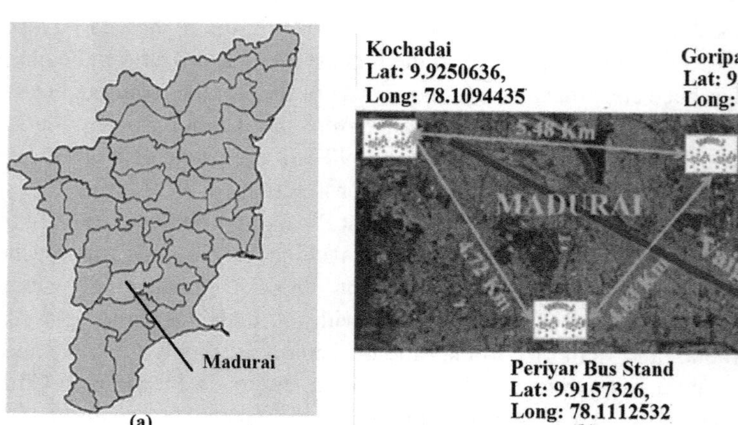

FIGURE 7.1 Data assortment at three different geolocations.

The particulate monitor employs an inexpensive PMS 3003 laser particle sensor with a 0.3–10 μm detection range and 10 seconds response time [13]. Air is drawn by a fan through a chamber, where it is exposed to a laser beam, and a 90° scattered beam is detected by a photo-diode [14]. The wavelength of the laser source was not available from the manufacturer, but it was measured to be 675 nm by using a variable bandwidth Lambda 35 spectrophotometer model 365 (PerkinElmer, Inc.) [15]. Using a solution of 60.06 mg/L potassium dichromate, this instrument's absorbance accuracy was tested at 235, 257, 313, and 350 nm. Every approximate pulse height is directly proportional to the particle size, according to the Mie principle, and thus the scattered light is transformed into $PM_{1.0}$, $PM_{2.5}$ concentrations [16]. In addition to sampling, surface meteorological data such as temperature, wind speed, wind direction, and relative humidity were also measured simultaneously [17].

7.2.2 Data Analysis

The generated baseline data were statistically analyzed by variance analysis (ANOVA), Pearson correlation, and MLR analysis methods [18]. ANOVA was carried out in one direction 19 to investigate the geographical and temporal (weekly) variability of the pollutant concentrations. The ANOVA test findings demonstrated the degree of variability in the study's data. An ANOVA test compares the variances (2) of more than two groups to see if the means (*i*) are identical. The *F*-distribution can be used to achieve this. The *F*-test measures the proportion of variation that may be ascribed to error or unmeasured variance. Equation (7.1) serves as the null hypothesis against which the test is run [19].

$$H_0 : \mu_1 = \mu_2 = \cdots = \mu_i \tag{7.1}$$

H_1: not all μ are equal.

In the null hypothesis (H_0), it is initially presumed that there are no substantial variations between the means of the various populations. If the estimated *F*-value is found to be lower than the 5% meaningful level of the *F*-value table, if the alternative hypothesis (H_1) is not accepted, the null hypothesis will be accepted [20]. To analyze the relationship between the levels of the contaminants and the prevailing meteorological parameters, MLR and PC analysis were also carried out. All the statistical analysis was performed using the Microsoft Excel package.

7.2.3 Multilinear Regression

The statistical regression approach (MLR) is used to determine the association between the concentrations of ambient $PM_{1.0}$ and $PM_{2.5}$. Making assumptions, choosing plausible approaches, interpreting the results, and formulating the model are all possible starting points for MLR formulation [21,22]. Stepwise regression is a trustworthy technique to utilize when dealing with statistically stable data, which is why it is used in this study. Constructing the formula using the assumptions (see Equation 7.1) and its derivatives (see Equation 7.2).

$$y = B_0 + B_1X_1 + \cdots + B_nX_n + \varepsilon \tag{7.2}$$

where y is the projected value of the contemporary variable; B_0 i the y-intercept (value of y when all other parameters are set to 0); B_1X_1 is the regression coefficient (B_1) of the first independent variable (X_1); B_nX_n is the regression coefficient of the last independent variable; and ε is the formulation error.

7.2.4 Pearson's Correlation

Karl Pearson proposed and introduced the PC, and the area of science and engineering is advised to use it [23,24]. Equation has been used to determine the association between the $PM_{1.0}$ and $PM_{2.5}$ concentrations at each sampling location.

The linearity between the $PM_{1.0}$ and $PM_{2.5}$ makes sense toward the relationship among each-other to understand their values in one-another respect.

$$\text{sim}_i = \frac{\Sigma_{j=1}^{N}\left(X_j - \bar{X}\right)\left(Y_{ij} - \bar{Y}\right)}{\left(\sqrt{\Sigma_{j=1}^{N}\left(X_i - \bar{X}\right)^2}\right) - \left(\sqrt{\Sigma_{j=1}^{N}\left(Y_{ij} - \bar{Y}\right)^2}\right)} \tag{7.3}$$

where X_j and Y_{ij} are the sample-scale attribute data for the project being evaluated and the i-th sample, respectively, and sim_i is the "fitting of the i-th sample" of the project dataset. Additionally, "i-th" is a sample of the project data set, and X and Y are the average values of the project scale attribute data that are anticipated to be examined.

7.3 RESULTS AND DISCUSSION

The ambient $PM_{1.0}$ and $PM_{2.5}$ concentrations were measured using a laser sensor-based portable particulate monitor. The statistical results on the pollutant levels for the sampling sites have been presented in Table 7.1.

The variation of hourly pollutant levels along with meteorological parameters at each sampling site has been presented in Figure 7.2.

TABLE 7.1
Descriptive Statistics for $PM_{1.0}$ and $PM_{2.5}$ Levels at the Sampling Locations

	Goripalayam		Periyar Bus Stand		Kochadai	
Descriptive statistics	PM_1 (µg/m^{-3})	$PM_{2.5}$ (µg/m^{-3})	PM_1 (µg/m^{-3})	$PM_{2.5}$ (µg/m^{-3})	PM_1 (µg/m^{-3})	$PM_{2.5}$ (µg/m^{-3})
Mean	33.61	42.17	38.72	47.18	28.02	32.56
Median	32	39	38	47	27	31
Mode	29	34	37	48	31	31
Standard deviation	10.46	12.29	11.09	13.72	8.02	8.74
Kurtosis	−0.11	0.93	0.51	1.05	−0.11	−0.31
Skewness	0.60	0.81	0.23	0.65	0.57	0.49
Range	48	70	57	71	38	40
Minimum	15	20	12	19	14	17
Maximum	63	90	69	90	52	57

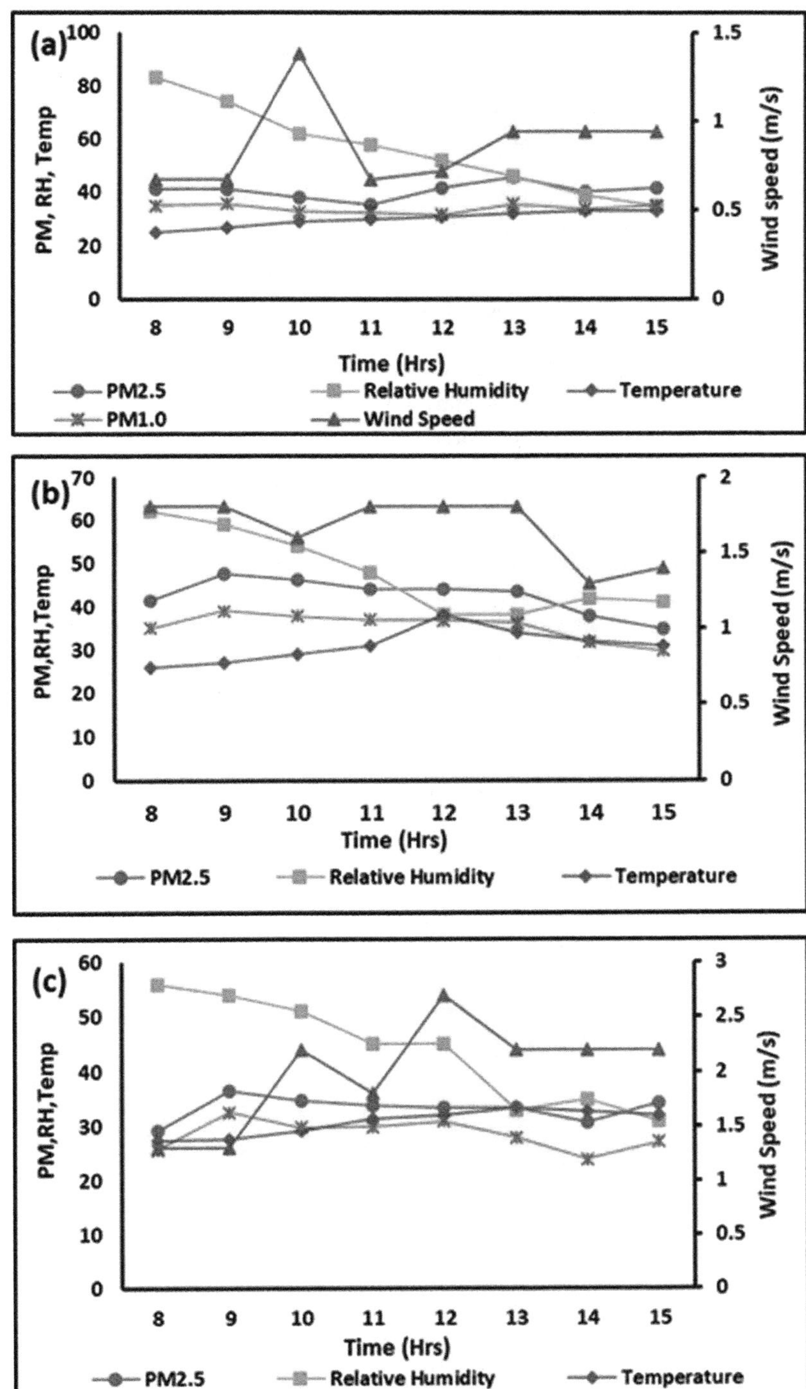

FIGURE 7.2 Hourly variation of average PM levels and meteorological parameters at (a) Goripalayam, (b) Periyar Bus Stand, and (c) Kochadai.

At Goripalayam, $PM_{1.0}$ and $PM_{2.5}$ levels were measured during the research period to range from 15 to 63 g/m^{-3} and 20 to 90 g/m^{-3}, respectively, whereas at Periyar Bus Stand, the same values ranged from 12 to 69 g/m^{-3} and 19 to 90 g/m^{-3}, respectively. At Kochadai, the $PM_{1.0}$ and $PM_{2.5}$ concentrations ranged between 14 and 52 μg/m^{-3} and 17 and 57 μg/m$^{-3,}$ respectively.

7.3.1 Hourly Variations in the Particulate Levels

At Goripalayam, the $PM_{1.0}$ and $PM_{2.5}$ concentrations were high at 8 am and 2 pm, and low at 11 am (Figure 7.2), whereas, at Periyar Bus Stand, higher concentrations were recorded at 9 am and 1 pm, and lower concentrations were recorded at 3 pm (Figure 7.2). At Kochadai, the concentrations were high at 9 am and low at 2 pm (Figure 7.2). The morning peak concentrations (between 8 and 9 am) observed during the sampling periods could be attributed to the higher vehicular activities since most of the personal, school, and college vehicles would pass through the sampling sites during that time only [3]. The afternoon peak values could be due to the wind-generated resuspended road dust since mostly higher wind speeds were recorded during that time [4]. Concentrations after morning are higher, the pollutant levels tend to decrease toward noontime. This could be attributed to the variation in the boundary layer height (BLH) with time [22]. The BLH had a significant influence on the pollutant levels until noon or evening. In the study area, the temperature and wind speed increased after sunrise, which accompanied the formation of the boundary layer and resulted in the thickening of the layer around noontime [23]. The increase in BLH provided higher volume and eased the dilution of the pollutants resulting in a decrease in particulate levels [24]. This could well be justified by the obtained negative correlation relationship between the pollutant levels and temperature (Table 7.2).

In addition to that, a reduction in vehicular activities could also be accounted for the reduced concentrations at that time. At Kochadai, the pollutant levels remained almost the same during the sampling hours. Kochadai is an industrial area, and two significant industries were being operated, and the emission rate of pollutants from these two sources would almost be considered invariant. So the sustained pollutant concentrations could mainly be attributed to these industrial emissions in addition to other sources.

In order to assess whether or not there was a substantial difference in pollutant levels between the sampling sites as well as between the sampling weeks, ANOVA was used in one direction, and the findings were presented in Table 7.3.

TABLE 7.2
Correlation of Meteorological Parameters and PM Concentrations

	Goripalayam		Periyar Bus Stand		Kochadai	
	$PM_{1.0}$	$PM_{2.5}$	$PM_{1.0}$	$PM_{2.5}$	$PM_{1.0}$	$PM_{2.5}$
Wind speed	−0.49	0.05	0.8	0.74	−0.05	0.07
Relative humidity	0.11	−0.21	0.42	0.41	0.44	0.03
Temperature	−0.16	−0.19	−0.1	−0.25	−0.31	−0.07

TABLE 7.3
ANOVA Results for the Temporal and Spatial Variability of the PM Levels

		$PM_{1.0}$		$PM_{2.5}$	
		F-value	*p*-value	*F*-value	*p*-value
Temporal variability[a]	Goripalayam	0.21	0.99	0.24	0.97
	Periyar Bus Stand	0.48	0.86	0.59	0.75
	Kochadai	0.37	0.77	0.49	0.83
Spatial variability[b]	–	0.98	0.44	0.39	0.72

[a] *F*-value is 2.24 (temporal variability);
[b] *F*-value is 2.64 (spatial variability).

TABLE 7.4
Meteorological Conditions Prevailed During the Sampling Hours

	Goripalayam			Periyar Bus Stand			Kochadai		
	Min	Max	Mean	Min	Max	Mean	Min	Max	Mean
Wind speed (m/s)	0.2	1.38	0.9	0.4	1.8	1.7	1.3	2.7	1.98
Relative humidity (%)	35	83	56.1	38	62	47.7	31	56	43.7
Temperature (°C)	23	34	30	26	38	31	27.5	33.3	30.7

It was observed from the results that during the study period, there was no substantial differences found at the 5% significance stage in both the $PM_{1.0}$ and $PM_{2.5}$ fractions. This revealed the fact that irrespective of the nature of the sampling sites, the city was polluted uniformly by the pollutants of current concern in the entire study period.

Even though there were mild variations in the pollutant levels during the sampling hours, there were no significant variations in the pollutant levels observed at all three sampling sites, and the results depicted that Madurai city was uniformly polluted by the $PM_{1.0}$ and $PM_{2.5}$ fractions. These small variations could be attributed to the variation in the emission sources such as traffic, windblown resuspended road dust, and prevailing meteorological conditions during the sampling hours.

7.3.2 Influence of Meteorological Parameters on PM Levels

The influence of the local meteorological parameters on particulate levels was studied by focusing on hourly wind speed, relative humidity, and temperature values. The summary of the observed meteorological conditions has been presented in Table 7.4. Wind speed is the lowest side in terms of Min, Max, and Mean value at Goripalayam station; and at Kochadai station, it is the highest for the same terms (see Table 7.4). However, in the case of relative humidity, it is the opposite that is highest at Goripalayam, and the lowest at Kochadai (see Table 7.4). And, at Periyar Bus Stand, both wind speed and relative humidity are medium among the three stations; and in terms of temperature, it is highest at the station.

TABLE 7.5
MLR Analysis Results for PM Concentrations

	Goripalayam		Periyar Bus Stand		Kochadai	
	$PM_{1.0}$	$PM_{2.5}$	$PM_{1.0}$	$PM_{2.5}$	$PM_{1.0}$	$PM_{2.5}$
Intercept	111.7	93.6	−9.7	−20.7	10.2	56.14
Wind speed	−2.8	−0.11	−8.1	−9.38	1.66	1.95
Relative humidity	−0.33	−0.26	−0.29	−0.43	−0.22	−0.09
Temperature	−1.88	−1.28	−0.57	−0.86	−0.18	−0.74

The relationship between the pollutant levels and meteorological parameters could be explained by correlating them individually, whereas, while all these dominating parameters influence the pollutants simultaneously, which is the real scenario, the prediction becomes more complex. In this case, the MLR model has been proved to be a handy tool to establish a much better relationship between the pollutant concentrations and the influencing meteorological parameters, since, in this model, all the predictors are being applied simultaneously and have been shown much better agreement. In this perspective, by taking the particulate levels as a dependent parameter, MLR analysis was performed. The meteorological values, as independent parameters, such as wind speed, relative humidity, and temperature values, and the results obtained were summarized in Table 7.5.

The results showed that at Goripalayam and Periyar Bus Stand, temperature, wind speed, and relative humidity all negatively correlated with the concentrations of $PM_{1.0}$ and $PM_{2.5}$, whereas at Kochadai, wind speed positively correlated with both concentrations of $PM_{1.0}$ and $PM_{2.5}$ and negatively correlated with both temperature and relative humidity. Among the meteorological parameters, the wind speed had a significant influence on the pollutant levels except on $PM_{2.5}$ concentrations at Goripalayam, where the temperature had a major influence. Wind speed has a vital role in dispersion of PM; it can also be enlightening up to 26%–63% of the variation in the relative exposure concentration (REC). By using these climatic variables, policies and urban design techniques to lessen bicycle exposure pollution might be evaluated more effectively [25]. Also, intercept represents the mean value of the response variable, hence, those show the highest relation at Goripalayam, followed by Kochadai station. Moreover, significantly the negative intercept appeared at Periyar Bust Station that extends the regression line downwards until it reaches the point where it crosses the dependent variables, overestimating on an average the dependent values, thereby a possibility of a negative correlation in the independent, and the highest correlation between both PMs are appeared.

The influence of temperature on particulate concentrations, which is complex in nature, has been reported in plenty of earlier studies worldwide. The change in temperature and wind, on the one hand, affects the vertical mixing, which results in the dilution of the pollutants in the boundary layer [26]. This vertical mixing is stronger in hot seasons as the temperature is also high. On the other hand, the photochemical formation of particulates is highly influenced by temperature. Even though the higher temperature helps the photochemical reaction to form more particulate pollutants in the atmosphere, when the temperature is too high, the pollutants tend to be in their gaseous states themselves [27]. In the present study, the recorded temperatures during the sampling periods were found to be very high. Hence, the negative relationship between the temperature and particulate levels could be understood by the elevated dilution processes and so the reduced pollutant levels. The wind speed and the relative humidity also had a negative relationship with the pollutant levels, which could be attributed to the dispersion and atmospheric scavenging effects of the wind and relative humidity, respectively [28]. The average relative humidity values were found to be either mostly below 62% or sometimes above 83% at Goripalayam and mostly below 62% at Periyar Bus Stand and Kochadai. The scavenging effect might have dominated over the hygroscopic growth of particulate pollutants since only the relative humidity range of 62%–83% favors the accumulation of fine particulates in the atmosphere [29].

7.3.3 Correlation of $PM_{1.0}$ with $PM_{2.5}$

The correlation between the $PM_{1.0}$ and $PM_{2.5}$ levels at each sampling site has been shown in Figures 7.3–7.5. It was found from the results that $PM_{1.0}$ levels clearly correlated with $PM_{2.5}$ levels at all the sampling sites.

The $PM_{1.0}/PM_{2.5}$ ratios were also calculated and are presented in Figure 7.6. It was clearly observed from the results that at least 80% of the $PM_{2.5}$ fraction was composed of $PM_{1.0}$ fraction. The obtained Pearson correlation coefficients between them, which were 0.59, 0.9, and 0.82 at Goripalayam, Periyar Bus Stand, and Kochadai, respectively, also confirmed this fact. The results can be acceptable if the R^2 values

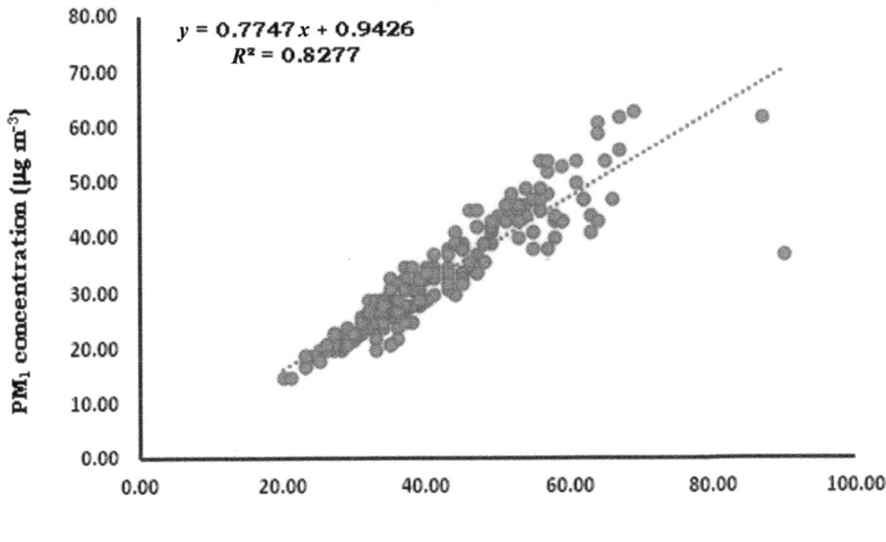

FIGURE 7.3 Correlations between $PM_{1.0}$ and $PM_{2.5}$ at Goripalayam.

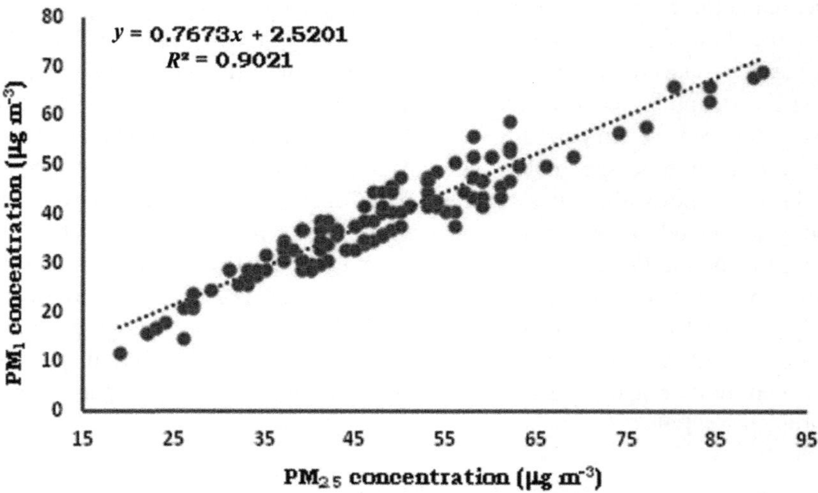

FIGURE 7.4 Correlations between $PM_{1.0}$ and $PM_{2.5}$ at Periyar Bus Stand.

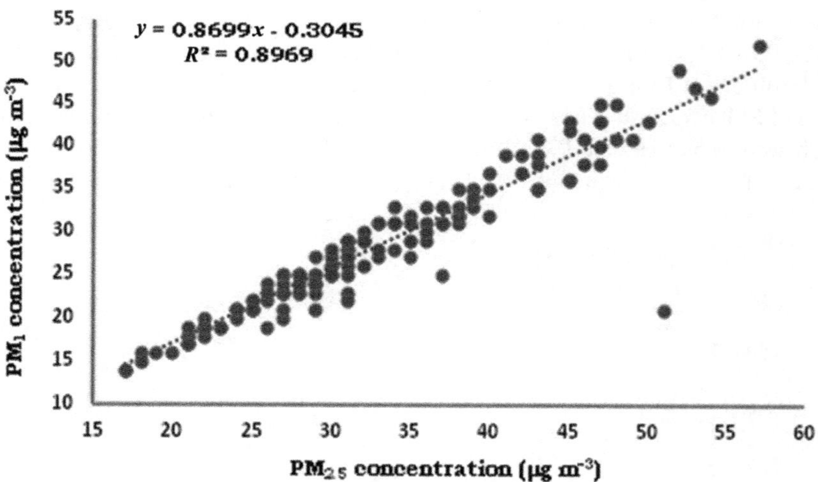

FIGURE 7.5 Correlations between $PM_{1.0}$ and $PM_{2.5}$ at Kochadai.

are about 0.5 or above. In the present study, the obtained R^2 values of the regression relation between $PM_{1.0}$ and $PM_{2.5}$ were above 0.82 at all the sampling sites. So, it could be concluded that the major portion of the $PM_{2.5}$ fraction was composed of $PM_{1.0}$ fraction in the study locations.

7.3.4 Standard Exceedances

The $PM_{2.5}$ data assorted at the three sampling sites were analyzed for the standard exceedances by comparing it with the 24 hours. Average National Ambient Air Quality Standards (NAAQS) (India) and WHO standards which are 60 and

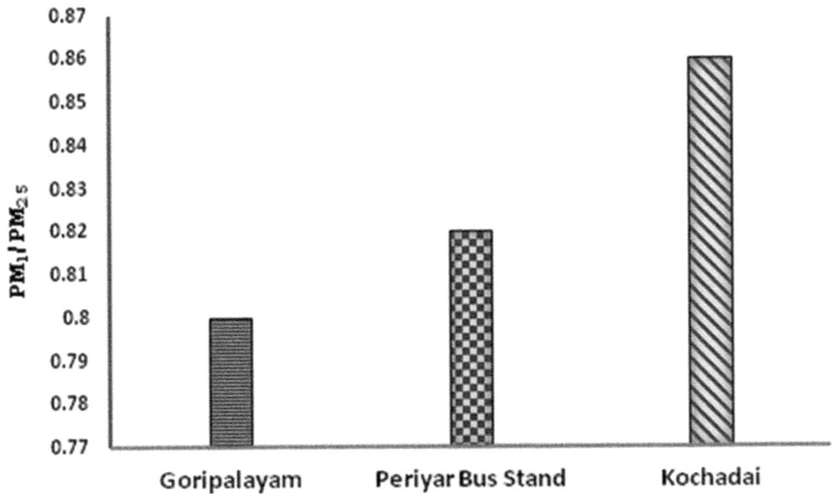

FIGURE 7.6 Average $PM_{1.0}/PM_{2.5}$ ratios at the sampling sites.

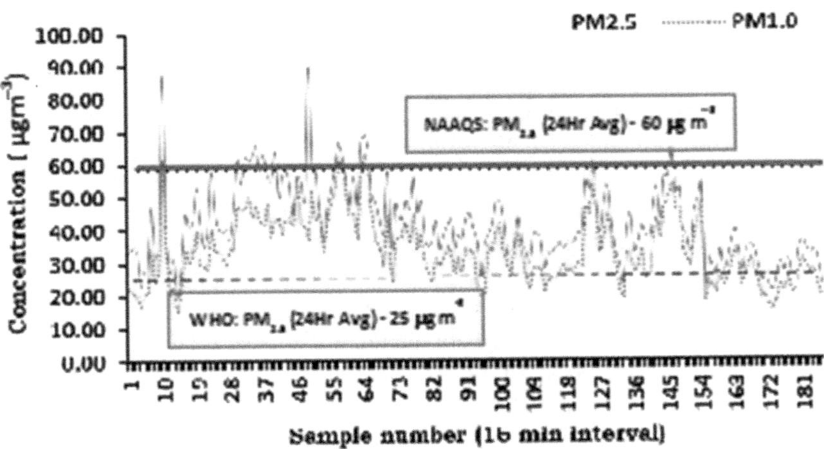

FIGURE 7.7 Variation of PM levels showing standard exceedances at Goripalayam.

25 μg/m^{-3}, respectively (applicable for industrial, residential, and rural areas, which includes traffic sites also), and the results have been shown in Figures 7.7 and 7.8.

It was revealed from the analysis that 96%, 95%, and 82% of the $PM_{2.5}$ concentrations recorded during the sampling hours exceeded the WHO standards at Goripalayam, Periyar Bus Stand, and Kochadai, respectively, whereas 9%, 14% of the recorded $PM_{2.5}$ concentrations exceeded the NAAQS values at Goripalayam and Periyar Bus Stand, respectively. This indicated that immediate preventive measures along with regular assessment with reference to this pollutant would be mandatory for maintaining a sustainable environment in the near future.

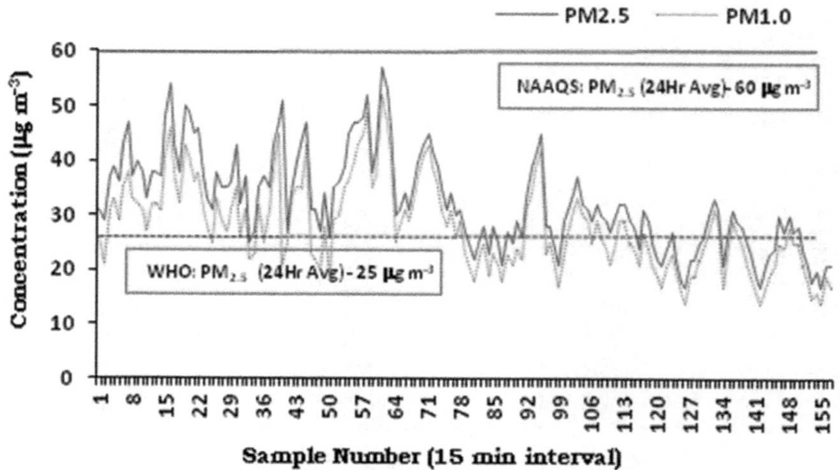

FIGURE 7.8 Variation of PM levels showing standard exceedances at Periyar Bus Stand.

7.4 CONCLUSION

For the first time, the present study provided a preliminary assessment of the concentrations of $PM_{1.0}$ and $PM_{2.5}$ in the urban environment of Madurai city. The study recorded a marked hourly variation in the PM levels, and it was attributed to the boundary layer parameters. The early morning and afternoon peak concentrations recorded could be attributed to the increased vehicular activities and windblown resuspended dust, respectively. Statistical results suggested that temperature, wind speed, and relative humidity negatively correlated with the PM levels at Goripalayam and Periyar Bus Stand, whereas, at Kochadai wind speed, it had a positive correlation. The results also suggested that wind speed significantly influenced the PM levels except for $PM_{2.5}$ at Goripalayam, where the temperature significantly impacted that pollutant. It was observed from the study that at least 82% of the assorted $PM_{2.5}$ data exceeded the WHO (24 hours average) standards during the study period. Hence, it could be concluded that regular and periodic monitoring of these pollutants and immediate strict preventive measures would be mandatory from the perspective of epidemiology.

REFERENCES

1. Harding, S. *et al.* Association of ambient air pollution with blood pressure in adolescence: a systematic-review and meta-analysis. *Curr. Probl. Cardiol.* 48, 101460 (2022).
2. Hawkins, T. W. & Holland, L. A. Synoptic and local weather conditions associated with $PM_{2.5}$ concentration in Carlisle, Pennsylvania. *Middle States Geogr.* 43, 72–84 (2010).
3. Du, Y., Xu, X., Chu, M., Guo, Y. & Wang, J. Air particulate matter and cardiovascular disease: the epidemiological, biomedical and clinical evidence. *J. Thorac. Dis.* 8, E8 (2016).

4. Srimuruganandam, B. & Nagendra, S. M. S. Analysis and interpretation of particulate matter–$PM_{1.0}$, $PM_{2.5}$ and PM_1 emissions from the heterogeneous traffic near an urban roadway. *Atmos. Pollut. Res.* 1, 184–194 (2010).
5. Singh, V., Biswal, A., Kesarkar, A. P., Mor, S. & Ravindra, K. High resolution vehicular PM_{10} emissions over megacity Delhi: relative contributions of exhaust and non-exhaust sources. *Sci. Total Environ.* 699, 134273 (2020).
6. Chen, T. *et al.* Acute respiratory response to individual particle exposure ($PM_{1.0}$, $PM_{2.5}$ and PM_{10}) in the elderly with and without chronic respiratory diseases. *Environ. Pollut.* 271, 116329 (2021).
7. Zhang, J. J. Y. *et al.* Predicting intraurban airborne $PM_{1.0}$-trace elements in a port city: land use regression by ordinary least squares and a machine learning algorithm. *Sci. Total Environ.* 806, 150149 (2022).
8. Liu, B. *et al.* Characterization and spatial source apportionments of ambient PM_{10} and $PM_{2.5}$ during the heating period in Tian'jin, China. *Aerosol Air Qual. Res.* 20, 1–13 (2020).
9. Luo, L. *et al.* Fine particulate matter ($PM_{2.5}/PM_{1.0}$) in Beijing, China: variations and chemical compositions as well as sources. *J. Environ. Sci.* 121, 187–198 (2022).
10. Ramkumar, M. *et al.* Neolithic cultural sites and extreme climate related channel avulsion: evidence from the Vaigai River Basin, southern India. *J. Archaeol. Sci. Reports* 40, 103204 (2021).
11. Ramasamy, M., Nagan, S. & Kumar, P. S. A case study of flood frequency analysis by intercomparison of graphical linear log-regression method and Gumbel's analytical method in the Vaigai river basin of Tamil Nadu, India. *Chemosphere* 286, 131571 (2022).
12. Zhang, Y. *et al.* Short-term effects of ambient PM_1 and $PM_{2.5}$ air pollution on hospital admission for respiratory diseases: case-crossover evidence from Shenzhen, China. *Int. J. Hyg. Environ. Health* 224, 113418 (2020).
13. Qian, L. *et al.* Toxic effects of boscalid in adult zebrafish (Danio rerio) on carbohydrate and lipid metabolism. *Environ. Pollut.* 247, 775–782 (2019).
14. Galvão, A., Aleixo, M., De Pablo, H., Lopes, C. & Raimundo, J. Microplastics in wastewater: microfiber emissions from common household laundry. *Environ. Sci. Pollut. Res.* 27, 26643–26649 (2020).
15. Verma, T. N., Sahu, A. K. & Sinha, S. L. Numerical simulation of air pollution control in hospital. *Air Pollut. Control* 2018, 185–206 (2018).
16. Querol, X. *et al.* PM_{10} and $PM_{2.5}$ source apportionment in the Barcelona Metropolitan area, Catalonia, Spain. *Atmos. Environ.* 35, 6407–6419 (2001).
17. Yin, Z. & Wang, H. Role of atmospheric circulations in haze pollution in December 2016. *Atmos. Chem. Phys.* 17, 11673–11681 (2017).
18. Majewski, G. *et al.* Concentration, chemical composition and origin of PM_1: results from the first long-term measurement campaign in Warsaw (Poland). *Aerosol Air Qual. Res.* 18, 636–654 (2018).
19. Ostro, B. *et al.* Long-term exposure to constituents of fine particulate air pollution and mortality: results from the California teachers study. *Environ. Health Perspect* 118, 363–369 (2010).
20. Sivaramasundaram, K. & Muthusubramanian, P. A preliminary assessment of PM_{10} and TSP concentrations in Tuticorin, India. *Air Qual. Atmos. Heal.* 3, 95–102 (2010).
21. Bhagat, S. K., Tung, T. M. & Yaseen, Z. M. Heavy metal contamination prediction using ensemble model: case study of bay sedimentation, Australia. *J. Hazard. Mater.* 403, 123492 (2020).
22. Krishnaraj, R. Control of pollution emitted by foundries. *Environ. Chem. Lett.* 13, 149–156 (2015).

23. Krishnaraj, R. Contemporary and futuristic views of pollution control devices in foundries. *Ecotoxicol. Environ. Saf.* 120, 130–135 (2015).
24. Krishnaraj, R. Investigation of sand filtration techniques to reduce secondary pollution in wet scrubber. *Eur. J. Sci. Res. Rev.* 89, 384–393 (2012).
25. Hu, H. *et al.* Field investigation for ambient wind speed and direction effects exposure of cyclists to $PM_{2.5}$ and PM_{10} in urban street environments. *Build. Environ.* 223, 109483 (2022).
26. Elminir, H. K. Dependence of urban air pollutants on meteorology. *Sci. Total Environ.* 350, 225–237 (2005).
27. Krishnaraj, R., Sakthivel, M. & Devadasan, S. R. Performance efficiency of wet scrubber in induction furnace towards green revolution- a case study in Indian foundry. *J. Environ. Res. Dev.* 6, 824–833 (2012).
28. Krishnaraj, P. Green revolution for sustainable development in Indian industries using CFD analysis. *J. Environ. Res. Dev.* 6, 802–813 (2012).
29. Krishnaraj, R. Foundry air pollution: hazards, measurements and control. In: Lichtfouse, E., Schwarzbauer, J., Robert, D. (eds) *CO_2 Sequestration, Biofuels and Depollution* (335–357). Springer, Cham (2015).

8 Innovations Via CFD
Studies on Pollution Control Equipment Design and Multiphase Flow Simulation

S. Venkatesh, S. Vijayan, and M.M. Matheswaran

8.1 INTRODUCTION

The CFD is a vital tool in the design and optimization of pollution-protecting equipment. In order to analyze fluid flow, heat transfer, and chemical reactions in these pollution control systems, CFD simulations have become essential tools. Without the need for costly and time-consuming physical prototyping, CFD can offer important information on the behavior and performance of various pollution control systems by employing numerical methods to solve the governing equations. By harnessing the power of CFD, engineers can now develop state-of-the-art pollution control devices that not only meet stringent environmental standards, but also increase industrial efficiency and sustainability. In this study, a widely used pollution control device such as cyclone separator's performance and efficiency have been discussed. Moreover, the CFD simulation that provides engineers and scientists with a powerful tool to analyze fluid dynamics and optimize their designs for a variety of applications from pollution control to transportation and more. Some key applications include: Air pollution control equipment: CFD is used to study the flow and dispersion of pollutants in industrial emission sources, allowing engineers to optimize the design and performance of scrubbers, cyclones, electrostatic precipitators, and fabric filters. Moreover, CFD is helpful to analyze the energy consumption to operate the pollution control equipment in industries. During the optimization of emission control systems geometries, experts consider energy consumption because altering the geometry increases or decreases the energy required to operate the systems. Water Pollution Control Systems: In water treatment plants and industrial water treatment plants, CFD helps to understand flow dynamics, mixing and pollutant removal efficiency, leading to better design of clarifies, aeration basins, and other treatment processes. Noise control devices: CFD is used to analyze the acoustic behavior of noise control devices such as mufflers and mufflers, allowing engineers to optimize their designs for effective noise reduction. Traffic Emissions Management: CFD is used to study engine combustion processes and emissions, facilitating the development of cleaner, more fuel-efficient power systems for vehicles, aircrafts, and ships.

DOI: 10.1201/9781003452072-8

8.2 FUNDAMENTALS OF CFD

Different velocity fields in turbulent flow lead to variations in conveyed attributes, including species concentration, energy, and momentum. In practical engineering calculations, direct simulation is computationally expensive due to the turbulent flow's high-frequency and small-scale variations. To overcome this challenge, methods such as time averaging or ensemble averaging can be used to manipulate the instantaneous governing equations and effectively filter small scales. This modification adds unknown variables that require the use of turbulence models to represent them in known quantities. The main characteristic of eddy currents is the different nature of their velocity fields. These techniques efficiently remove small scales, leading to altered equations that require less computing power. Generally, three types of turbulence schemes have been applied by researchers in past decades for simulating the pollution control equipment such as Reynolds Stress Model (RSM), k-ε model, and Large Eddy Simulation (LES). Predominantly, RSM was applied for analysis due to its accuracy (Fu et al., 1987). Few authors applied LES for simulation; it also produces significant results in pressure and particle trajectories. In CFD simulation, two types of flow schemes opted by the researchers, one is a steady-state scheme, and another one is the transient scheme. The accuracy of transient simulation is mostly dependent on the total flow time, number of time steps, CFL number, and time step size. Furthermore, the ultimate results selection heavily relies on the time-averaging procedure.

8.2.1 k-ε Model

The k-ε model requires the specification of certain input parameters or values to accurately model the turbulence behavior in a given flow scenario. It is a two-step equation model. It requires near-wall treatment in mesh generation for solving the viscous sublayer effects. Mostly, when designing cyclone separators, the viscosity-affected region was solved at the top of the vortex finder by many researchers. The recommended *Y*+ value for these simulations is 30–200 (Fluent, 2004). Gong et al. (2012) investigated the effects of leaf margins and helix angle on the internal flow field of the cyclone. In that work, the axial flow cyclone was used for simulation. The RNG k-ε model was applied for simulation in that study (Fluent, 2004). In that study, the multiphase flow was solved by a stochastic trajectory model with one-way coupling. Some of the most important parameters/settings of the k-ε model are:

Intensity of Turbulence (I): The degree of turbulence in a flow is determined by its intensity. The root mean square (RMS) ratio between the fluctuations in turbulent velocity and the mean flow velocity is the standard definition. Fluent (2004) recommends a turbulence intensity ratio of 5% for the cyclone separator simulation.

Turbulent Viscosity Ratio: This represents the liquid's molecular viscosity (μ) divided by the turbulent viscosity. It explains how turbulence affects the flow. Turbulent kinetic energy and damping are used to compute turbulent viscosity in the k-ε model. Fluent (2004) recommended the turbulence viscosity ratio for the cyclone separator simulation in 10.

Length Scale (L): Characteristic length scale used in the model, often related to the size of turbulent eddies. In some variants of the k-ε model, this length scale is estimated from local flow properties (Fluent, 2004).

Wall Characteristics: The existence of a wall influences the turbulence behavior in areas close to walls. Eddy flow close to walls is modeled using wall functions instead of independently solving the viscous sublayer. Conditions at the outset and the boundary: At the start of the simulation, the initial conditions for the Turbulence Kinetic Energy (TKE) and Dissipation Rate (DR) of turbulence must be established. Furthermore, according to Fluent (2004), these variables need to be subject to suitable boundary conditions at the inlet, outflow, and other domain boundaries.

Solution Settings: Solution settings such as convergence criteria, time step size (simulations), and numerical discretionary methods (finite volume methods) are important for accurate and stable simulations.

Turbulence Prandtl Numbers: They are used to establish a relationship between TKE and DR and eddy viscosity and diffusivity. Equations for turbulence: The transport equations such as TKE and DR should be solved in the k-ε model. According to Fluent (2004), these equations depict how turbulence develops in the flow domain.

User Experience Factor: Some variants of the k-ε model may have user-defined constants that affect the behavior of the model and may be adapted to better fit experimental data or specific flow conditions (Fluent, 2004).

A small number of writers used k-ε to simulate cyclones in the past. Azadi et al. (2010) investigated three different cyclone separator sizes using the CFD method. The k-ε model was used to examine the impact of turbulent modeling. The study found that with variances of 2.3%, 3.4%, and 3.6% of the experimental data for cyclones I, II, and III, respectively. This methodology accurately predicted the cut-off diameter. A Fluent-based CFD model was used to simulate gas-particle flow inside a cyclone. The simulation findings validated the effectiveness of CFD modeling with RSM as a feasible strategy for investigating the effect of cyclone size on performance measures.

Shin et al. (2005) inspected the design and numerical and experimental aspects of building high-efficacy cyclones. The k-ε model with two equations is utilized to solve Reynolds stresses. In a Lagrangian frame, the particle trajectory computation includes drag, centrifugal, and Coriolis forces. Further evidence indicates that increasing the temperature and pressure typically has a noteworthy outcome on the competence of fine particle collection.

Ma et al. (2000) examined the flow pattern in small-scale cyclones by CFD. The study assumes that the presence of particles has no effect on fluid flow in the cyclone due to the minimal particle loads that are typically found in most small air sampling cyclone applications. The governing equations are solved using the RNG turbulent model. According to the study's findings, the fluid flow is not axially symmetrical, the axis' computational treatment is critical, and the results are heavily reliant on it. However, by accounting for the fluid flow's time dependency, the prediction of particle penetration may be more accurate. This will have an impact on numerical forecasts for microscopic particles.

8.2.2 Reynolds Stress Turbulence Model

The RSM is often used in CFD simulations for more complex flows where isotropic turbulence effects are important (Fluent, 2004). It is a seven step equation model. It produces more accuracy in the cyclone separator and venturi scrubber simulations compared to k-ε model. However, high mesh refinement and validation are required for this model. Previously, many researchers applied this RSM for cyclone separator simulation.

Reynolds Stress Transport Equations: Additional transport equations for each of the Reynolds stress tensor's (RST) constituent parts must be solved for RSM Turbulence's interaction with mean flow and turbulent fluctuations is often taken into consideration via six extra equations (for 3D simulations).

Boundary Conditions: Appropriate boundary conditions for the properties of the RST must be defined at the domain boundaries. Wall functions similar to the k-ε model can also be used to describe turbulence behavior near walls (Fluent, 2004).

Initial Conditions: The initial conditions for the properties of the RST must be set at the beginning of the simulation. Solver Settings: Numerical settings of the solver, such as discretization models, time step size (simulations), and convergence criteria, are critical to obtaining accurate and stable results (Fluent, 2004).

Turbulence Prandtl Numbers: These are used to relate eddy viscosity and diffusivity to components of the Reynolds stress tensor. Treatment of walls: Proper treatment of turbulent flow near walls is essential. Some RSM applications use low-Reynolds-number wall treatments or transformations to capture near-wall behavior (Fluent, 2004).

Model Constants: RSM often contains more model constants compared to simpler models such as k-ε. In order to obtain accurate results under certain flow conditions, some of these constants must be determined from experimental data or calibration.

Solver Approach: RSM simulations can be computationally more demanding due to the increased number of equations and variables. Ensuring convergence and stability may require careful tuning of solution settings (Su et al., 2011).

Geometry and Flow Type: RSM is frequently applied to intricate flows, including separated flows, eddies, and strongly isotropic flows. The complexity of the flow and the availability of experimental data for model calibration should be taken into consideration when deciding whether to utilize RSM.

Mesh Quality: A well-structured and sufficiently fine mesh is crucial to accurately capture the turbulence behavior. It is especially important to set the mesh near the wall.

Post-Processing: RSM simulations can provide more detailed information about turbulence behavior such as isotropic effects and Reynolds stress distribution. Post-processing tools must be used to analyze and visualize simulation results. Many authors applied RSM in cyclone simulation in past decades.

Venkatesh et al. (2020a,b) used RSM to analyze the square cyclone separator's performance. The kinetic energy and momentum of the turbulence were solved in that simulation using the Second Order Upwind (SOU) method. Using the SIMPLEC approach, the pressure-velocity coupling was handled. Venkatesh et al. (2021a,b) constructed three sequentially connected square cyclones. With a 2.2 μm particle size, this cyclone in series is predicted to have a 61% collection efficacy. Its collection efficiency is 10.2% higher than that of a single square cyclone. When the particle size is at least 10 µm, it has been shown that exceptional gathering efficiency is achieved by connecting three square cyclone separators in series. A series design also results in a 14.3% smaller pressure drop when compared to a single square cyclone. The pressure drop decreases as one moves from the first to the last square cyclone. The results showed that cyclone (A) collects larger particles, whereas cyclones B and C collect smaller ones, according to both experimental and flow pattern analysis.

Bernardo et al. (2006) used a 3D CFD approach to characterize the gas and the gas-solid flow in cyclones. That examination concludes that when there are particulates in the flow, the scroll inlet section angle proposal led to an underestimate of the gas velocity. By using the RSM, new turbulence and tangential velocity characteristics were obtained. While the traditional intake duct achieved an overall collection efficiency of 54.4% under the same operating conditions, the 45° inlet section angle increased that percentage to 77.2%, as shown in the numerical findings.

Through CFD research, Venkatesh et al. (2019a,b) adjusted the geometry of the cyclone separator. It has been discovered that key inlet dimensions help to improve cyclone separator performance by reducing pressure loss (Venkatesh et al., 2019a,b). In addition, the velocity profiles suggest that the tangential velocity is lower than that of a typical cyclone. Additionally, the new design's axial velocity is lower than that of the conventional cyclone. It indicates that there is less particle exiting through the outlet port. It suggests that the new design has high collection efficiency. In that research, authors developed radial profiles through CFD results for analyzing the pressure and velocity in a deep manner, which is shown in Figure 8.1. Additionally, the radial profile demonstrates that as the radial position increases, so does the velocity. Its value declines in the outer region after reaching its maximum.

Alt text: Figure 8.1 indicates velocity in different radial positions from the central axis. The maximum velocity was shown in red color and the minimum velocity was indicated in blue color. Intermediate velocity was indicated in yellow and green colors. Based on these results, collection performance is estimated in cyclones.

Elsayed and Lacor (2012) computationally examined four cyclone separators with different dust output geometries: without a dustbin, with a dustbin, with a dustbin plus a dip leg, and with a dip leg. This study's primary goal was to measure the air cyclones' flow field pattern and performance. Ultimately, the findings indicate that the four investigated cyclones had nearly identical maximum tangential velocities. There is no radial acceleration in the area around the cyclone. Here, the cyclone separator was modeled using the RSM, and particle trajectories were executed by the Discrete Phase Model (DPM).

The multi-cyclone arrangement in series and parallel was studied by Venkatesh et al. (2020a,b). The chemical and aerosol sampling industries utilize multi-cyclones to extract micron-sized particles from gas or air medium. In that work, four cylindrical cyclones with optimum dimensions were connected in parallel. The findings of that study imply that the dustbin has a square shape. Similar to circular designs, this shape increases wall friction, which prevents the gas from flowing freely inside the trash. It also contributes to the gas' decreased velocity. Similar to how the gas exits the cylinder section of the cyclone, it quickly contracts at the cone section.

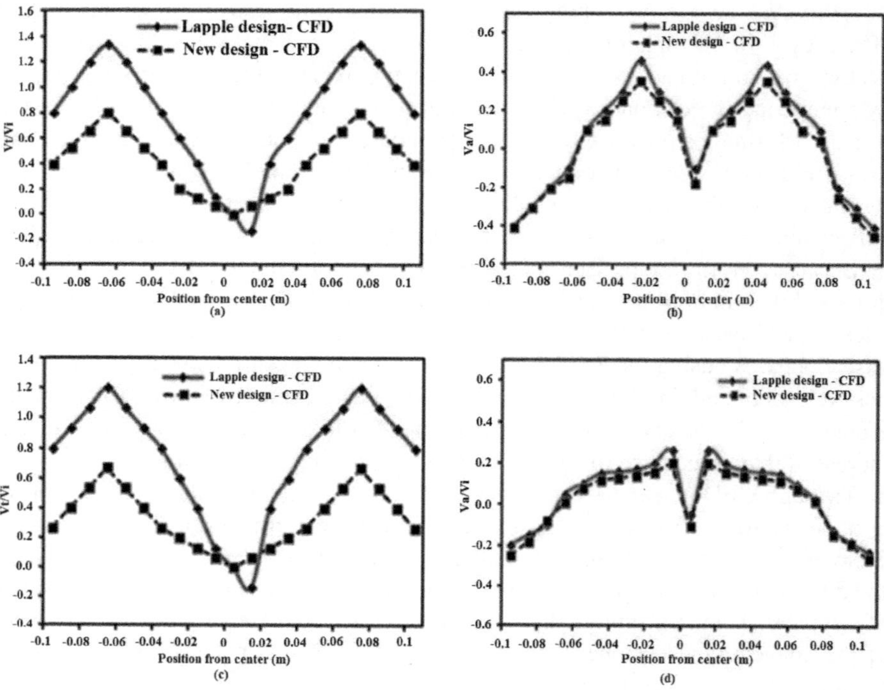

FIGURE 8.1 Radial profiles- velocity (Venkatesh et al., 2019a,b).

It is further compressed along the tube's increased length. Then, as the gas flow is directed downhill, the gas is abruptly enlarged at the dustbin's square shape. The gas suddenly contracts and expands, reducing the velocity inside the cyclone. Each cyclone experiences a smaller pressure drop as a result of this reduced velocity.

In order to simulate the venturi inlet cyclone separator mathematically, certain geometrical features were taken into account (Venkatesh and Sakthivel, 2017). The length, breadth, and outlet diameter of the venturi inlet were among the most crucial geometrical parameters optimized. According to Venkatesh and Sakthivel (2017), after fitting the quadratic polynomial equation using the response surface approach, the quadratic model's relevance was evaluated using ANOVA. The research indicates that the region of the venturi intake wall, which is red in both the new and mathematical designs, is the location with the highest pressure. In addition, the lowest pressure is observed at the vortex detector and the blue output of the cyclone separator in both the new and mathematical models. Consequently, the pressure at the exit of the cyclone separator is lower than that at the venturi inlet wall region. Radial velocity is greatest at the vortex finder zone and lowest at the venturi inlet region, according to the contour plots for both setups. In addition, the vortex finder zone is where the axial and tangential velocities are at their highest, while the inlet region is where they are at their lowest. Furthermore, compared to the mathematical model, the three velocities of the optimized model are substantially lower.

Chemical engineers employ bottom intake cyclone separators with venturi to extract particles from air or gas media (Venkatesh et al., 2019a,b). To improve cyclone performance, important geometrical parameters were improved, such as venturi inlet width, cyclone body height, and cyclone total height. The central composite design approach was used to fit the regression equation using response surface methods. According to the results of that research, the pressure drop reaches its highest value at the venturi area for newly constructed cyclones that are mathematically and ideally optimized. The pressure drop is minimal at the outlet area of both kinds. The findings of the flow pattern analysis show that the collection bin of the optimally designed cyclone contains more particles than the mathematical model predicted. In addition, the math model shows that the optimal design cyclone's conical section has the most particles spinning around it.

Su (2006) looked into how well cyclone separators worked with various intake layouts. The various types of inlet conditions were modeled for simulation in that study. The RSM was applied in that simulation to find the effects of various inlets. Lastly, an analysis was done on how different inlet configurations affected the cyclone separator. The findings show that the double inlet of the cyclone produces a better particle flow pattern compared to others. Similarly, Elsayed and Lacor (2011) investigated various inlet conditions and their flow pattern status in cyclone separators by RSM. That study shows that with the exception of the core area, the axial velocity profiles of the three cyclones are quite similar. The impact of cyclone inlet width or height on tangential velocity is the most significant. The maximum tangential velocity decreases with increasing width or height of the cyclone entrance.

Venkatesh et al. (2021b) used RSM to modify the hydrocyclone's shape. In that simulation, from 0 (far from the wall) to 1 (near the wall), the functional magnitude varies. As a result, in opposing pressure gradients, that model yields more accurate findings. Moreover, the residual value of 10^{-4} establishes accurate convergence. Perceived pressure and velocity flow continue until they reach a constant. For the first iteration, hybrid initialization was carried out. The iteration is then started at 0.0001 seconds. There are 20 iterations for each time step and five overall time steps. To create the precise pressure field and compel mass communication, the SIMPLE algorithm was applied. To boost the simulation accuracy, the SOU and QUICK scheme was applied in that research.

8.2.3 LES Model

LES is a computational approach to CFD that aims to capture large-scale turbulent structures in a flow by modeling small-scale turbulence. Compared to traditional Reynolds Averaged Navier Stokes (RANS) models, it provides a more detailed and accurate representation of turbulence. Here are some key equipment settings and considerations to keep in mind when using an LES model in CFD simulations:

Subgrid-Scale (SGS) Model: LES requires a subgrid-scale model to account for the effects of unresolved small-scale turbulence. Commonly used SGS models include the Smagorinsky model, dynamic models, and Wall-Adapting Local Eddy viscosity (WALE) models.

Mesh Resolution: LES simulations require a finer mesh resolution compared to RANS simulations to accurately capture large-scale turbulent structures.

The grid should be structured and refined near regions and boundaries of interest.

Time Integration: LES simulations can be performed in both time-accurate (unsteady) and pseudo-constant modes. Time integration systems must be carefully selected to ensure stability and accuracy. Explicit time steps are often used. Boundary conditions: Proper use of boundary conditions is crucial. The inlet, wall, and outlet conditions must be well-defined to minimize the effect of numerical errors on the resolved turbulence.

Numerical Dispersion: In LES simulations, attention should be paid to controlling numerical dispersion. Advanced numerical methods or shock capture methods can be used to reduce overdispersion.

Filter Width: LES involves filtering the flow field to separate the resolved large scales from the unresolved small scales. The width of the filter should be chosen appropriately based on the desired level of resolution and the size of the turbulent structures in the flow.

Initial Conditions: Proper initial conditions, including velocity and turbulence fields, must be provided for the LES simulation.

Wall Modeling: LES requires accurate modeling of the area near the wall. Wall-resolved LES simulations aim to directly resolve near-wall structures, while wall-modeled LES simulations use wall functions or similar approaches to model near-wall turbulence behavior. Creating a turbulent inflow: In an unsteady LES simulation, it can be difficult to create proper turbulent inflow conditions. Synthetic methods or preliminary simulations can be used to generate suitable turbulent inflow profiles.

Solver Settings: Solver settings such as time step size, spatial discretization patterns, and convergence criteria must be carefully chosen to ensure accurate results and stable simulation.

Post-Processing: LES simulations produce a wealth of information that can provide detailed information about turbulence behavior. Advanced post-processing techniques are often required to analyze and visualize the results.

Validation and Verification: LES simulations should be validated with experimental data or other high-quality simulations whenever possible to ensure the accuracy and reliability of the results. Particularly for complicated geometries or high Reynolds number flows, LES simulations can be computationally intensive and need a large number of computer resources. To get accurate and significant results, rigorous validation and careful adherence to settings and parameters are necessary. In previous decades, very few authors used this LES to simulate the cyclone separator.

De Souza et al. (2012) obtained flow pattern results in small-scale cyclones through LES. The particle dynamics were computed through the Discrete Phase model (DPM.) The LES modeling has a benefit over RANS. It requires parameter adjustments unrelated to any particular set of operating conditions or cyclone geometry. The results of two distinct collection criteria and three alternative integration procedures for the particle motion equations are presented and discussed in that work. By comparing the computed and experimental grade efficiencies, it was possible to see how sensitive they were to shape and Reynolds number variations. Moreover, LES offers a robust and reliable modeling alternative for such flows.

Kumar et al. (2024) studied the elliptical cross-section cyclone separator by the LES model. In that work, the benchmark model for performance comparisons is the standard model, which has a circular cross-section ($D = 0.205$ m). The exact hydraulic diameter and, by extension, volume, of each cyclone has been assured. The study evaluated the processing vortex core's periodicity around the cyclone axis, pressure drop, and collecting efficiency using a rapid Fourier transform on the temporal velocity signal. Results consistently demonstrate that reducing pressure losses and boosting collection efficiency are achieved by raising the major axis values. As the elliptical cross-section grows longer, the components of mean velocity change more rapidly than the components of velocity fluctuations. To remove larger particles with minimal pressure drop, pre-separators with the largest major axis models are used.

Guo et al. (2023) studied how a cylinder vortex stabilizer affects the fluctuating turbulence structure of a Stairmand cyclone separator by the LES approach. Any current cyclone model can use the simple cylinder vortex stabilizer without requiring a significant replacement. It is believed that this novel cyclone body modification will mitigate the negative impacts of ash hopper particle entrapment and vortex end swing in turbulent flow. Three new models with different vortex stabilizer lengths and diameters were incorporated in the numerical simulations that were based on the Stairmand cyclone separator. Based on the findings, it appears that the cylinder vortex stabilizer has the potential to increase flow instability and improve fluctuating turbulence structure.

Pandey and Brar (2023) studied the effects of bulged conical sections through LES. Using the two fixed ends of the straight cone wall, the curved profile was used in its place to create a cyclone model that resembles a pot. Then, five distinct cyclone configurations were analyzed with curvature radii of 1.5, 1.25, 1.0, 0.75, and 0.5 m. The LES approach has been utilized to assess pressure losses, collecting efficiency, and the mean and variable flow field. Conclusive findings show that as the curvature size increases, so does the pressure drop and the collecting efficiency. The suggested model may be used as the pre-separator since it has a bigger separation space than the traditional design, allowing it to accumulate more solid particles.

8.3 PARTICLE DYNAMICS

Settings in DPM within CFD simulations can significantly influence the accuracy and the behavior of particle tracking. The choice of settings are completely depends on nature of particle's simulation. It's essential to thoroughly understand the behavior of particles in the system and validate DPM settings to ensure accurate and meaningful results.

DPM is commonly used in CFD for simulating the behavior of particles, droplets, or bubbles in various applications, including spray combustion, environmental pollution dispersion, and pharmaceutical manufacturing processes (Fatahian et al., 2020). Moreover, it is highly utilized in cyclone separators for predicting particle trajectories and capturing efficiency (Li et al., 2015; Wang et al., 2020; Wei et al., 2020). In cyclone separator simulation, two kinds of particle simulation are carried out, one is uniform-size particle simulation and another one is the Rosin Rammler approach for non-uniform-size particle simulation. Here are some essential settings and parameters to consider when using DPM.

8.3.1 Particle Properties

- Particle Diameter: Define the diameter of particles being tracked. It's crucial for collision and drag force calculations (Morsi and Alexander, 1972).
- Particle Density: Specify the density of particles relative to the fluid density.
- Particle Inertia: Set the particle inertia relative to the fluid inertia.

8.3.2 Injection Settings

- Injection Type: Choose the method of particle injection, such as continuous or discrete.
- Injection Location: Define where particles are injected into the domain.
- Particle Mass Flow Rate: Specify the rate at which particles are injected.
- Particle Velocity: Set the initial velocity of injected particles.

8.3.3 Particle Tracking

- Lagrangian Tracking: Decide how particles should be tracked through the flow field.
- Integration Time Step: Determine the time step used to update particle positions.
- Stochastic Sub-Models: Consider enabling sub-models for particle dispersion, coalescence, and breakup if applicable.

8.3.4 Boundary Conditions

- Particle-Wall Interaction: Define how particles interact with walls (e.g., reflect, stick, or rebound).
- Inlet/Outlet Conditions: Specify conditions at inlet and outlet boundaries for particles.
- Interactions with Fluid
- Drag Coefficient Model: Choose a model for calculating drag forces on particles (e.g., Stokes, Schiller-Naumann, etc.).
- Lift Force Models: Enable lift force models for particles if needed.
- Virtual Mass Force: Consider including virtual mass forces if applicable.
- Collision Models
- Particle-Particle Collisions: Specify collision models if multiple particles can interact.
- Coalescence and Breakup: Set parameters for particle coalescence and breakup if relevant to your simulation.

8.3.5 Output and Visualization

- Data Sampling: Decide when and where to collect data about particle behavior (e.g., at specific locations or time intervals).
- Visualization: Choose how to visualize particle trajectories, concentrations, or other relevant data.

8.3.6 Solver Controls

- Time Step Control: Ensure that the time step used for DPM tracking is appropriate, often linked to the CFL condition.

8.4 METHODS FOR IMPROVE SIMULATION ACCURACY

Improving simulation accuracy in CFD involves a combination of careful model selection, proper tuning, accurate numerical methods, and experimental data validation. Here are some methods and strategies to improve simulation accuracy.

Mesh Refinement: Use a finer mesh to better define flow characteristics and gradients. Refinement of the mesh near boundaries, areas of interest, and areas with strong gradients can lead to more accurate results. The quality of mesh has been validated by three methods, namely, Grid Independence Study (GIS), Mesh Metric Evaluation (MME), and Grid Convergence Index (GCI) study. In MME, the grid aspect ratio is calculated by taking the ratio of fine mesh to medium mesh and medium mesh to coarse mesh. The mesh aspect ratio should be above 1.3 (Slack et al., 2000). Moreover, the Jacobian ratio values are calculated for validation. In GIS, the result of the simulation is validated against each set of mesh generated. Because, sometimes, the grid size affects the simulation results. Therefore, the optimum grid size is to be predicted by this study for getting accurate results. Another accurate method for validating the mesh quality is GCI. In this study, the grid refinement ratio is evaluated by obtaining different kinds of mesh elements. Afterward, the asymptotic value is computed by standard equations. The asymptotic value is close to 1, and then the mesh quality is superior (Roach et al., 1986). Use AMR techniques to dynamically refine or thin a mesh based on local flow characteristics. This ensures efficient use of computing resources while maintaining accuracy.

Turbulence Models: Select the appropriate turbulence model based on the flow characteristics. Moving from simpler models (e.g., RANS) to more complex ones (e.g., LES) as needed to accurately capture turbulence behavior. Wall treatments: Choose wall treatments suitable for areas along the wall. Advanced wall modeling can help better predict wall-bounded flows. In recent research, RSM and LES models have been mostly utilized by authors in cyclone simulation. For these turbulence models, accurate mesh refinement is required to produce accurate results. Therefore, viscosity-affected regions should be properly solved by additional grid generation near the wall. Further, the $Y+$ values should computed by this treatment. It is helpful for predicting the accurate time step size. It reduces the unnecessary simulation time and error.

Higher-Order Numerical Schemes: Use higher-order numerical discretization schemes (e.g., quadratic or higher) for spatial and temporal derivatives to reduce numerical errors and improve accuracy. Previously, many authors applied SIMPLE, SIMPLEC, Pressure Implicit with splitting of operators (PISO), QUICK, and SOU schemes were applied in cyclone separator simulation. Predominantly, the SOU scheme was applied for TKE and DR. Further, the SIMPLEC and PISO scheme was applied in cylindrical cyclone simulation. The QUICK scheme was applied in the square cyclone simulation. The abovementioned schemes provide accurate results in pressure and velocity coupling.

Validation: Compare simulation results with experimental data or analytical solutions to confirm accuracy. If necessary, adjust model parameters and settings to match experimental trends. Previously, Su and Mao (2006) developed an experimental setup for square cyclone analysis. Moreover, they validated CFD results against this experimental data. It provides a satisfactory agreement. Therefore, many authors applied this procedure for validation of CFD codes in square cyclones (Safikhani et al., 2011; Venkatesh et al., 2020a,b; Fatahian et al., 2020). The validation results are shown in Figure 8.2. In cylindrical cyclone simulation, Elsayed and Lacor (2010) validated the results of Hoekstra (2000).

Alt text: Figure 8.2 represents the validation of CFD codes against experimental results in square cyclone separators in previous research. In this validation plot, the pressure drop results from the experimental analysis are utilized to validate the CFD simulation results. This validation result proves that CFD codes exactly match with experimental results with minimum deviation.

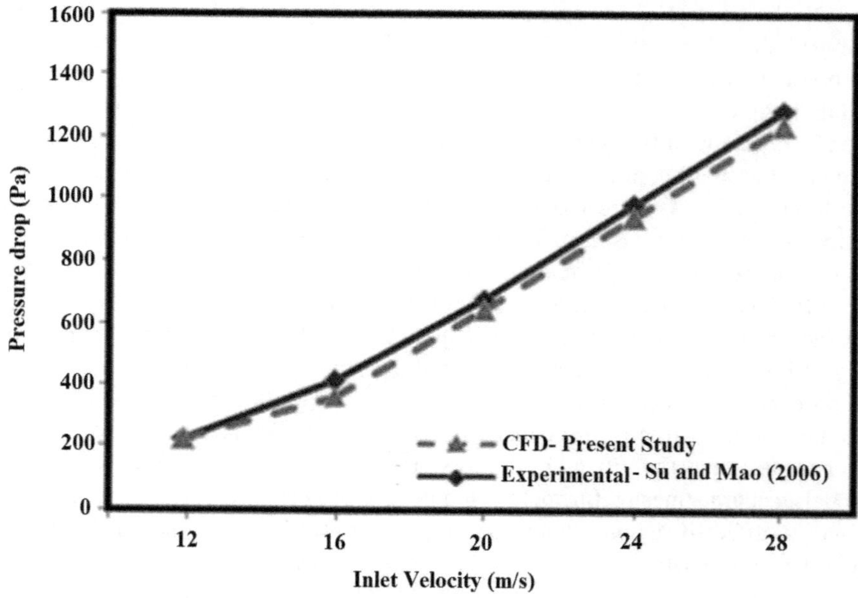

FIGURE 8.2 Validation results (Safikhani et al., 2011; Venkatesh et al., 2020a,b; Fatahian et al., 2020).

Physical Models and Solver Settings: Include more physical phenomena such as heat transfer, multiphase interactions, or chemical reactions to capture the full flow behavior. Carefully choose solver settings such as time step size, convergence criteria, and stabilization methods to ensure stability and accuracy. The CFL number is an important one in time step size selection. It is a dimensionless number used in CFD to assess the accuracy and stability of numerical simulations of fluid flow. It is defined as:

$$\mathrm{CFL} = \frac{u \times \Delta t}{\Delta x} \tag{8.1}$$

Where u is the characteristic velocity of the flow, Δt is the time step used in the simulation, and Δx is the characteristic spatial grid size.

A CFL number less than 1 is typically required for stability in CFD simulations. If the CFL number is much smaller than 1, it may indicate that the time step used is overly conservative, potentially leading to longer simulation times. On the other hand, a CFL number close to 1 suggests that the time step is appropriately chosen for the given flow conditions. It's essential to adjust the time step and grid size in CFD simulations to maintain a stable and accurate solution, as the CFL number can affect the convergence and reliability of the results. After the selection of the proper time step size, provide suitable iterations per time step size.

Numerical Loss Control: Controls numerical loss using methods that minimize dispersion or dispersion. This can help maintain sharp slopes and flow characteristics.

Model Calibration: Calibrate model constants or coefficients using experimental data to improve the fit between simulation and actual behavior.

Boundary Conditions: Use accurate boundary conditions that represent the actual physical conditions of the problem. Consider both inflow and outflow conditions. In previous research, most of the authors applied inlet velocity and pressure outlet conditions for simulation in cyclones. Furthermore, in particle dynamics simulation, trap, escape, and reflect boundary conditions were opted according to the simulation conditions (Safikhani et al., 2011; Venkatesh et al., 2020a,b; Fatahian et al., 2020). For obtaining accuracy in simulation, particle density should be exactly mentioned as per the experimental analysis. In previous research, magnesium particles, chalk dust, and fly ash particles were utilized for analysis. Moreover, setting the step length factor plays a crucial role in the accuracy of particle simulation. Increasing the step length factor increases the accuracy. At the same time, increasing the abovementioned parameter required high computational time. Another important parameter, the maximum number of time steps factor provides the accuracy in simulation.

Convergence Monitoring: Monitor convergence behavior during simulations to ensure the stability and accuracy of solutions. If you experience convergence problems, adjust the resolution settings. To increase the accuracy of the simulation, many authors set 10^{-4} convergence criteria in cyclone separator simulation. Moreover, during the transient simulation, monitoring the solution convergence and contour conditions is important for increasing the accuracy.

Symmetry and Symmetry Conditions: Use symmetry conditions when necessary to reduce computing power and improve accuracy.

Parallel Computing: Leverage parallel computing resources to speed up simulations, enabling higher resolution and better accuracy.

Verification and Validation: Follow established verification and validation practices to ensure the correct application of numerical methods and models and the accuracy of results.

Post-Processing: Use appropriate post-processing techniques to analyze and visualize simulation results and identify areas for improvement.

User Knowledge: Develop knowledge of the CFD methods, techniques, and physics of a given problem to make informed decisions and optimize simulation accuracy.

Literature and Community Resources: Keep up to date with the latest research, publications, and community discussions related to CFD methods and best practices.

8.5 CONCLUSION

This study concludes that the CFD techniques and its different schemes are utilized in the cyclone separator and pollution control equipment for simulation. This study indicates that RSM and LES are mostly applied turbulence schemes in cyclone separators for simulation. Further, the SIMPLEC and PISO algorithms are predominantly applied for spatial discretization. In addition, this study provided various methodologies available for mesh quality inspection to obtain accurate results in simulation. Moreover, this study explains the numerical simulation settings required for proper simulation in cyclone separators. The requirement of *Y* plus values and its impact on simulation is provided in this study. The effect of CFL number and time step size selection method has been provided in this study.

REFERENCES

Azadi, M., Azadi, M., & Mohebbi, A. (2010). A CFD study of the effect of cyclone size on its performance parameters. *Journal of Hazardous Materials, 182*(1–3), 835–841.

Bernardo, S., Mori, M., Peres, A. P., & Dionisio, R. P. (2006). 3-D computational fluid dynamics for gas and gas-particle flows in a cyclone with different inlet section angles. *Powder Technology, 162*(3), 190–200.

De Souza, F. J., de Vasconcelos Salvo, R., & de Moro Martins, D. A. (2012). Large eddy simulation of the gas–particle flow in cyclone separators. *Separation and Purification Technology, 94,* 61–70.

Elsayed, K., & Lacor, C. (2010). Optimization of the cyclone separator geometry for minimum pressure drop using mathematical models and CFD simulations. *Chemical Engineering Science, 65*(22), 6048–6058.

Elsayed, K., & Lacor, C. (2011). The effect of cyclone inlet dimensions on the flow pattern and performance. *Applied Mathematical Modelling, 35*(4), 1952–1968.

Elsayed, K., & Lacor, C. (2012). The effect of the dust outlet geometry on the performance and hydrodynamics of gas cyclones. *Computers & Fluids, 68,* 134–147.

Fatahian, H., Hosseini, E., & Fatahian, E. (2020). CFD simulation of a novel design of square cyclone with dual-inverse cone. *Advanced Powder Technology, 31*(4), 1748–1758.

Fluent, Inc. (2004). *Fluent 6.1.22 Users' Guide.* Fluent Inc, Lebanon, NH.

Fu, S, Launder, B.E., & Leschziner, M.A. (1987). Modeling strongly swirling recirculating jet flow with Reynolds-stress transport closures. *Sixth Symposium on Turbulent Shear Flows.* Toulouse, France.

Gong, G., Yang, Z., & Zhu, S. (2012). Numerical investigation of the effect of helix angle and leaf margin on the flow pattern and the performance of the axial flow cyclone separator. *Applied Mathematical Modelling*, *36*(8), 3916–3930.

Guo, M., Lu, Y., Xue, C., Sun, X., & Yoon, J. Y. (2023). The impact of cylinder vortex stabilizer on fluctuating turbulence characteristics of a cyclone separator based on large eddy simulation. *Advanced Powder Technology*, *34*(9), 104149.

Hoekstra, A. J. (2000). Gas flow field and collection efficiency of cyclone separators. TU Delft, Ph. D. Thesis, Delft University of Technology.

Kumar, D., Jha, K., Kumar, V., & Brar, L. S. (2024). Performance evaluation of cyclone separators with elliptical cross-section using large-eddy simulation. *Powder Technology*, *438*, 119660.

Li, Q., Xu, W., Wang, J., &Jin, Y. (2015). Performance evaluation of a new cyclone separator–Part I experimental results. *Separation and Purification Technology*, *141*, 53–58.

Ma, L., Ingham, D. B., & Wen, X. (2000). Numerical modelling of the fluid and particle penetration through small sampling cyclones. *Journal of Aerosol Science*, *31*(9), 1097–1119.

Morsi, S. A. J., & Alexander, A. J. (1972). An investigation of particle trajectories in two-phase flow systems. *Journal of Fluid mechanics*, *55*(2), 193–208.

Pandey, S., & Brar, L. S. (2023). Performance analysis of cyclone separators with bulged conical segment using large-eddy simulation. *Powder Technology*, *425*, 118584.

Roache, P. J., Ghia, K. N., & White, F. M. (1986). Editorial policy statement on the control of numerical accuracy. *Journal of Fluids Engineering*, *108*(1), 2.

Safikhani, H., Akhavan-Behabadi, M. A., Nariman-Zadeh, N., & Abadi, M. M. (2011). Modeling and multi-objective optimization of square cyclones using CFD and neural networks. *Chemical Engineering Research and Design*, *89*(3), 301–309.

Shin, M. S., Kim, H. S., Jang, D. S., Chung, J. D., & Bohnet, M. (2005). A numerical and experimental study on a high efficiency cyclone dust separator for high temperature and pressurized environments. *Applied Thermal Engineering*, *25*(11–12), 1821–1835.

Slack, M. D., Prasad, R. O., Bakker, A., & Boysan, F. (2000). Advances in cyclone modelling using unstructured grids. *Chemical Engineering Research and Design*, *8*(78), 1098–1104.

Su, Y. (2006). The turbulent characteristics of the gas–solid suspension in a square cyclone separator. *Chemical Engineering Science*, *61*(5), 1395–1400.

Su, Y., & Mao, Y. (2006). Experimental study on the gas–solid suspension flow in a square cyclone separator. *Chemical Engineering Journal*, *121*(1), 51–58.

Su, Y., Zheng, A., & Zhao, B. (2011). Numerical simulation of effect of inlet configuration on square cyclone separator performance. *Powder Technology*, *210*(3), 293–303.

Venkatesh, S., & Sakthivel, M. (2017). Numerical investigation and optimization for performance analysis in Venturi inlet cyclone separator. *Desalination and Water Treatment*, *90*, 168–179

Venkatesh, S., Sakthivel, M., Avinasilingam, M., Gopalsamy, S., Arulkumar, E., & Devarajan, H. P. (2019a). Optimization and experimental investigation in bottom inlet cyclone separator for performance analysis. *Korean Journal of Chemical Engineering*, *36*, 929–941.

Venkatesh, S., Sakthivel, M., Sudhagar, S., & Daniel, S. A. A. (2019b). Modification of the cyclone separator geometry for improving the performance using Taguchi and CFD approach. *Particulate Science and Technology*, *37*(7), 795–804.

Venkatesh, S., Kumar, R. S., Sivapirakasam, S. P., Sakthivel, M., Venkatesh, D., & Arafath, S. Y. (2020a). Multi-objective optimization, experimental and CFD approach for performance analysis in square cyclone separator. *Powder Technology*, *371*, 115–129.

Venkatesh, S., Sakthivel, M., Saranav, H., Saravanan, N., Rathnakumar, M., & Santhosh, K. K. (2020b). Performance investigation of the combined series and parallel arrangement cyclone separator using experimental and CFD approach. *Powder Technology*, *361*, 1070–1080.

Venkatesh, S., Sivapirakasam, S. P., Sakthivel, M., Ganeshkumar, S., Prabhu, M. M., & Naveenkumar, M. (2021a). Experimental and numerical investigation in the series arrangement square cyclone separator. *Powder Technology*, *383*, 93–103.

Venkatesh, S., Sivapirakasam, S. P., Sakthivel, M., Krishnaraj, R., & Leta, T. J. (2021b). Investigation on hydrocyclone for increasing the performance by modification of geometrical parameters through cfd approach. *Desalination and Water Treatment*, *244*, 157–166.

Wang, Z., Sun, G., & Jiao, Y. (2020). Experimental study of large-scale single and double inlet cyclone separators with two types of vortex finder. *Chemical Engineering and Processing-Process Intensification*, *158*, 108188.

Wei, Q., Sun, G., &Gao, C. (2020). Numerical analysis of axial gas flow in cyclone separators with different vortex finder diameters and inlet dimensions. *Powder Technology*, *369*, 321–333.

9 A Geographic Information System Study of Gaseous and Particulate Pollutants in and around Tuticorin, India

Sivaramasundaram Krishnamoorthy, Nagaprasad Nagaraj, Shanmugam Ramaswamy, L. Priyanka Dwarampudi, Krishnaraj Ramaswamy, and S. Venkatesh

9.1 INTRODUCTION

The particulate matter (PM) contains liquid droplets and solid particles in the air. Usually, the PM is considered in two types, such as PM2.5 and PM 10. The PM 10 indicates that the solid particle diameter falls on 10 μm. Moreover, the PM2.5 indicates that the solid particle diameter falls on 2.5 μm. These PMs are easily inhalable. These PM sizes are measured through the particle counter. According to several studies, those gaseous pollutants and PM2.5 that can include important chemical components including sulfate, elemental carbon, and organic carbon are closely linked to negative health impacts [1]. However, those associations depend on the specific pollutants and geographic area. Tuticorin is one of the industrial cities in the state of Tamil Nadu, India. The land surface has erosion geomorphology. It is the host of one of the biggest thermal power plants in South India and one of the busiest harbors in the country. It is also the location of many chemical and fertilizer industries. This makes it one of the most significant coastal cities in India's southern Tamil Nadu region commercially. Additionally, these sectors have been causing environmental pollution for a very long time as a result of numerous human activities throughout the past several years. Due to this pollution, a lot of agricultural lands are spoiled. However, majority of the research based on water contamination of this area was conducted [2,3] and not in gaseous and particulate pollutants that led the motivation to conduct this significant research for the area.

DOI: 10.1201/9781003452072-9

A GIS technology combines standard database features like querying and statistical analysis with the advantages of spatial visualization and analysis offered by maps. GISs are unique among information systems because of these qualities, and they are helpful to a variety of public and private organizations for describing events, forecasting consequences, and formulating strategies [4–9]. The geographical distribution of natural resources in a geosystem is often determined using GIS.

The atmosphere, an integral part of geosystem, is one of the most important natural resources for the survival of life in this planet Earth, which is being harmfully destroyed due to the emission of pollutants from both natural and anthropogenic sources [8–11]. It is important to note that the analysis of air quality with reference to particulate pollutants is spatial, and the GISs are the tools that are specially meant for spatial analysis. This dynamic linkage will be one of the natural methods for solving the existing environmental pollution effectively, especially in urban areas [11–15]. So, the experimentally evaluated concentrations of the air pollutants of the present research work can be analyzed and displayed using this vital technology.

A few studies were created employing GIS for exposure assessment, characterization of PMs (PM10 and PM2.5) in a mining zone in India, and estimation of particle-gaseous pollutant emissions of woody biomass supply chain (BSC). According to authors, the dust is composed mostly of the following six key elements, accounting for 95% of the weight percentages: O, C, F, Si, Al, Fe, and Ca4. Additionally, the suggested model only reported on environmental emissions from transportation, and the rationale is because data for the harvesting and production processes in the woody BSC are not readily available [16]. One GIS study revealed the spatial variation of ambient PM load of Durgapur industrial town and surrounding areas [11]. Other GIS studies focused on visualization for forecasting of dust, PM (PM2.5 and PM10), gaseous pollutants (CO, O_3, SO_2, and NO_2), and the air quality index. Both studies valued appropriate techniques for reducing air pollution as well as GIS for online applications. On the years 2000–2003 and 2006–2008, a research team used the dispersion and chemical transport model EURAD (European Air Pollution Dispersion) to predict hourly concentrations of a number of pollutants on a horizontal grid in an urbanized region of Western Germany [17]. The model's ability to calculate the average exposure concentration for every location was praised; however, it can only predict unstable particles brought on by road activity. However, current study is unable to combine query and statistical analysis reports, and the proper visualization and geographic analysis were also missing. In addition to computation, economics was a constant focus of their studies.

With the use of GIS technologies, research may combine standard database operations like query and statistical analysis with the accurate visualization and spatial analysis that maps can only give. Numerous public and private organizations have benefited from the GIS's continuous unique characteristics for explaining events, forecasting results, and strategizing. When compared to more modern technologies, the GIS completes given work more quickly. Additionally, a GIS must be capable of data collection, administration, analysis, and modification as well as the presentation of results in both graphic and geographical data [18]. The ability of GISs to bridge the technological gap that exists among its need for researchers and decision-makers for simple understanding of the data is another significant aspect. The user-friendliness of GISs is a trait that has propelled GIS to the top of the list of most widely used planning platforms around the world. The capacity of GIS to address technical queries also makes it a valuable tool in many situations.

This research's objective was to explore the proposed area Tuticorin, India in terms of gaseous and PM through GIS and its constituents. Hence, authors studied (1) to describe seasonal distribution pattern of SO_X and NO_X, (2) spatial distribution in post-monsoon season and summer season (for SO_X and NO_X; and RSPM and TSP), (3) integrated seasonal for all; authors also presented, and (4) extents of distribution of the gaseous and particulate pollutants (after superimposition).

The section followed by study area, data collection, and GIS mapping methods in detail. Also, those presented graphical results described scientifically in results section with compressive discussion executed soon after. A conclusion set with the prospective studies.

9.2 METHODS

9.2.1 GIS Approach to Tuticorin Ambient Air Quality Studies

9.2.1.1 Study Area

One of the industrial cities in India's Tamil Nadu state is Tuticorin. It is physically situated close to the Gulf of Mannar at 8°48′N and 78°11′E latitude and longitude, respectively. It had 2,16,058 people according to the 2001 census, and it now has roughly 6,000 people. With hot, dry summers and rainy, moderate winters, the city experiences a hot, tropical climate. The temperature varies from 20°C to 35°C. High relative humidity between 60% and 75% prevails throughout the year. The land surface has erosional geomorphology. It is the host of one of the biggest thermal power plants in South India and one of the busiest harbors in the country. It is also the location of many chemical and fertilizer industries [19,20].

9.2.1.2 Data Collection

The air quality data were collected for six sampling sites in the city for present investigation. The sampling sites were selected in such a way that at least two sampling sites come under each of the sampling site categories, namely, residential, traffic–residential, and industrial areas. Out of the selected six sites, sampling was carried out for three sites and data from the State Pollution Control Board, Tamil Nadu was obtained for the remaining three sites. The air samples were collected using a commercially available respirable dust sampler (Envirotech APM 460 BL model), and the particulate levels were estimated by adopting United States Environmental Protection Agency (USEPA) recommended methods for gaseous pollutants and gravimetric methods for particulate pollutants.

9.2.1.3 Sampling Method

The respirable dust sampler used in this study is primarily made up of an inlet pipe, a cyclone separator, a brushless motor and blower enclosed in a shelter, and a filter adapter assembly with a gaseous sampling attachment. It can operate continuously around-the-clock with an input voltage ranging from 220 to 240 V. All of the parts that make up the sampler are protected from the elements by the shelter's thoughtful construction. An elapsed time meter that is connected in series with the blower measures the length of the sample process. A manometer that is connected to the device displays the air flow rate. It may be changed to get the appropriate air flow rate. A number of gaseous pollutants may be sampled using the respirable dust sampler, which includes a particular attachment for sampling them. Figure 9.1 shows an instrument used for sampling.

Typically, 1.1–1.4 cubic meters of air are extracted every minute. Two steps of filtering and cyclone separation are applied to the air inside the sampler. The larger particles (those between 10 and 100 microns in size) are collected in the first step using a cyclone separator. The remaining particles, which range in size from 0.1 to 10 microns, are gathered through a glass microfiber filter (Whatman GF/A) that has been dried and weighed beforehand. Therefore, the collection over the filter paper

FIGURE 9.1 Instrument used for sampling.

could represent the mass of PM10 (RSPM), and the collection within the container coupled to the cyclone separator might offer the mass of PM10-100. The particle pollution concentration has been estimated using a gravimetric determination process. The samples collected on filter sheets have not typically been weighed, despite the fact that the glass microfiber filters utilized for the collection of respirable particles have less affinity for moisture absorption. After conditioning for 48 hours at 25°C and a constant humidity of around 50%, the loaded and unloaded filters were weighed. By adding the concentrations of PM10 and PM10–100, the total suspended particulate (TSP) matter concentration has finally been determined.

9.2.2 GIS Mapping

The ERDAS IMAGINE 9.3 software was used for database coverage file preparation. The SOI toposheet (No. 58L/1, 58L/2 with the scale of 1:25,000) and a more current Tuticorin corporate map with 1:25,000 scales were used to create the base map of Tuticorin city. These maps were then inspected and approved by the Ministry of Defence, Government of India, and digitalized. The point data were created for the location of sampling stations through on-screen digitization. The errors in the point data were rectified by the clean/built commands for the creation of topology. Built and clean commands are for the creation of polygon topology and intersections, receptively [21]. These combinations give Virtual Worlds greater than the defaults displayed. The labeling was done for the location of the sampling stations. The coverage file was exported to shape the file format that would be suitable for ARC VIEW GIS software. The coordinates of the geographic latitude/longitude co-ordinate system were used to identify the locations of the air sampling sites, and the concentrations of the pollutants at the sample sites were obtained as another co-ordinate of the same system. The pollutants, TSP and RSPM, were categorized into five groups with different concentrations, and the recorded concentrations were evaluated. [22–24].

Database was imported from ERDAS IMAGINE 9.3 to ARC VIEW GIS 3.2a. The attribute table was prepared for sampling stations in order to link the nonspatial data (pollutant concentrations). Based on the attributes of the concentrations of the pollutants, their spatial distribution within the study area was done by using the spatial analysis module. By using the spatial interpolation method, isoline was developed for joining the places that had same concentrations. This process was done for all the pollutants including different seasons.

The pollutant concentrations were classified into five ranges as very high, high, medium, low, and very low. By using these ranges, their aerial extents were marked as polygon data. Different symbologies were given to show their ranges with suitable colors and patterns. The aerial extents were determined by using the script for area analysis. The final layout with a suitable title, legend, scale, etc. was developed, and the final thematic was generated. The methodology that was adopted for achieving the objectives is depicted in the following flow chart given in Figure 9.2 for a better understanding of the present work. The location of the sampling station and the location of sampling sites are illustrated in Figures 9.3a and b.

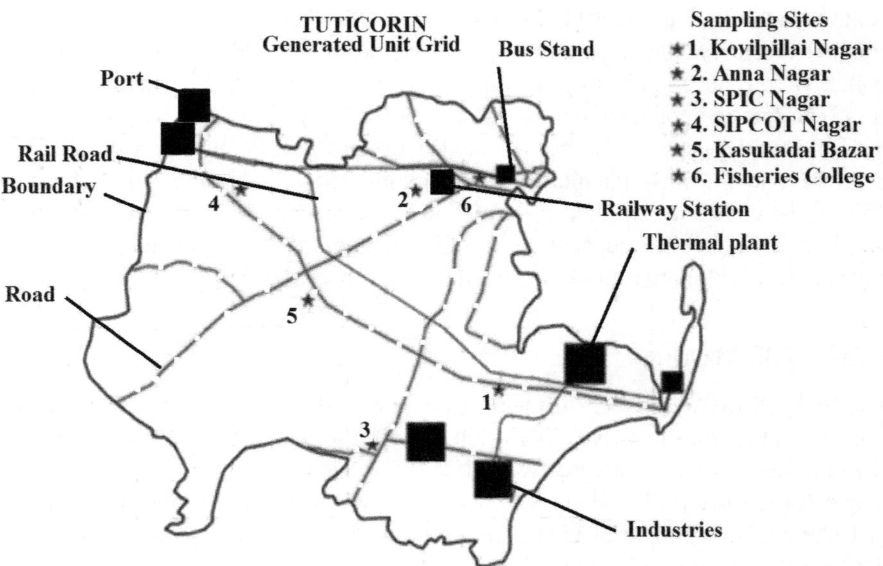

FIGURE 9.2 Flow chart depicting the processes carried out in developing a geographic information system (GIS) map.

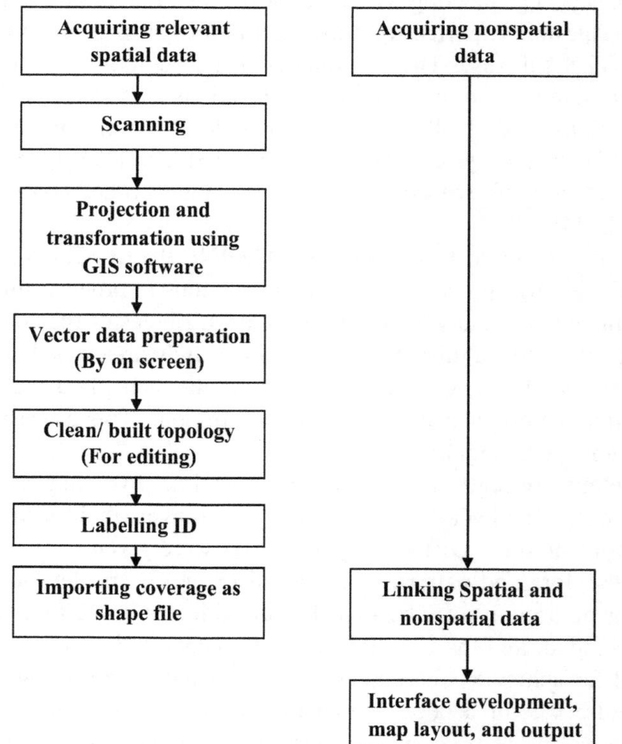

FIGURE 9.3 (a and b) Location of sampling station and location of sampling sites.

9.3 RESULTS AND DISCUSSIONS

With the aid of GIS technology, geographical distribution studies of gaseous pollutants like SO_2 and NO_x and particle pollutants like RSPM and TSP throughout all four seasons in Tuticorin are conducted in the current systematic study. With respect to the seasons, there is variation in the concentration ranges, which are between the minimum and maximum of the recorded pollutant concentrations. As already stated in the previous chapters, the meteorological conditions could determine the ambient air pollutant concentrations since emissions from the sources may be considered as constant all over the year in the city. So, the observed nonuniformity in the ranges of concentrations is mainly due to the meteorological conditions that were prevailing during the sampling period. Based on this fact, the observed pollutant concentrations of all the four seasons were divided into equal number of ranges but with different intervals, and the contours are developed, which may be useful for studying the season-wise spatial distribution pattern of the pollutants of present interest. The extents of distribution of various concentration ranges of the pollutants with respect to the seasons are calculated and have been presented in Tables 9.1–9.4.

GIS has the capacity to overlay several layers of geographically referenced data, allowing the user to see and analyze the relationships between diverse objects and structures in order to provide "new" information. Through this superimposing facility in GIS software, all the four seasonal themes of the pollutants are overlaid separately

TABLE 9.1
Seasonal Distribution Pattern of SO_2 in Tuticorin

Sl. No.	Season	Concentration Range (μg/m³)	Extent (Km²)
1	Post-monsoon (January – March)	< 4	51.42
		4–8	24.76
		8–12	17.24
		12–16	14.40
		> 16	5.31
2	Summer (April – June)	< 6	16.54
		6–12	42.55
		12– 18	24.27
		18–24	20.05
		> 24	9.76
3	Pre-monsoon (July – September)	< 6	21.42
		6–12	43.85
		12– 18	21.72
		18–24	17.21
		> 24	8.96
4	Monsoon (October – December)	< 6	12.13
		6–12	43.12
		12– 18	25.74
		18–24	21.21
		> 24	10.69

TABLE 9.2
Seasonal Distribution Pattern of NO_x in Tuticorin

Sl. No.	Season	Concentration Range (μg/m³)	Extent (Km²)
1	Post-monsoon (January – March)	< 12	51.42
		12–18	24.76
		18–24	17.24
		24–30	14.40
		> 30	5.31
2	Summer (April – June)	< 8	16.54
		8–12	14.54
		12–16	24.27
		16–20	20.05
		> 20	9.76
3	Pre-monsoon (July – September)	< 5	21.42
		5–10	43.85
		10–15	21.72
		15–20	17.21
		> 20	8.96
4	Monsoon (October – December)	< 12	12.39
		12–18	43.12
		18–24	25.74
		24–30	21.21
		> 30	10.69

for getting the composite themes, and the intersected ranges are eliminated for investigating the overall spatial distribution pattern of the pollutants properly [24–27].

The five ranges of concentrations in the seasonal data of NO_x and SO_2 are dissolved into three main ranges of concentrations such as high (>35 μg/m³), medium (30–35 μg/m³), and low (<30 μg/m³). Similarly, the five ranges of concentrations of RSPM and TSP are dissolved into three main ranges such as high (>275 μg/m³), medium (175–275 μg/m³), and low (<175 μg/m³) for reducing the complications and for the better understanding of the spatial distribution patterns of the pollutants by using the provisions in the GIS software. The calculated spatial dimensions of distribution in sq.km of the integrated maps that include all the seasons with high, medium, and low concentration areas for gaseous and particulate pollutants have been listed in Table 9.5. While the distribution of different concentration ranges of the pollutants of present interest with respect to the seasons such as post-monsoon, summer, pre-monsoon, and monsoon has been plotted and the maps of the year 2022 are displayed in Figures 9.4–9.8, the integrated maps that include the maps for all the seasons with high, medium, and low concentration areas are displayed in Figures 9.8–9.11, by which one can understand and interpret the ambient air quality with reference to the pollutants of present interest of Tuticorin easily [23–27].

Recent evidence indicates that the combined effect of pollutants from industries and urbanization adversely affects both biotic and abiotic systems. So, by adopting

TABLE 9.3
Seasonal Distribution Pattern of respirable suspended particulate matter (RSPM) in Tuticorin

Sl. No.	Season	Concentration Range (μg/m³)	Extent (Km²)
1	Post-monsoon (January – March)	< 15	18.99
		15–30	46.16
		30–45	30.62
		45–60	16.62
		> 60	0.77
2	Summer (April – June)	< 25	16.08
		25–50	39.23
		50–75	28.88
		75–100	23.76
		> 100	5.22
3	Pre-monsoon (July – September)	< 25	2.69
		25–50	16.60
		50–75	27.95
		75–100	35.09
		> 100	30.84
4	Monsoon (October – December)	< 25	8.28
		25–50	27.27
		50–75	38.98
		75–100	18.06
		> 100	20.56

the overlapping procedures, i.e., merging the attributes from the coverage, the final integration is carried out between the respective classes such as high, medium, and low in terms of the concentrations of the pollutants for getting new polygons, which can be identified as the areas that are having the combined effects [22–25]. These integrated maps are shown in Figures 9.8 and 9.10.

The systematic investigation on air quality reveals that the air pollutants of present interest are within the permissible limits, most of the times all over the year. So, it is essential to know the distribution pattern of the pollutants for different concentration ranges and the extents of their distribution. By this, one can not only identify the areas that are best suited for developmental activities but also the 'high concentration spots' for framing suitable abatement strategies and ultimately manage the atmosphere scientifically.

According to the current study, the monsoon season has the lowest SO_2 and NO_x concentrations (6.3 g/m^3 and 8.2 g/m^3, respectively), whereas the summer has the greatest values (125.3 g/m^3 and 129.7 g/m^3). The summer and post-monsoon seasons have the lowest concentrations of RSPM and TSP (17 g/m^3 and 31 g/m^3, respectively); whereas, the summer and pre-monsoon seasons have the greatest values (423 g/m^3 and 660 g/m^3, respectively). It is to be noted here that all other seasonal concentrations of the pollutants occur between these two extremes all over the study period.

TABLE 9.4
Seasonal Distribution Pattern of total suspended particulate matter (TSP) in Tuticorin

Sl. No.	Season	Concentration Range (μg/m³)	Extent (Km²)
1	Post-monsoon (January – March)	< 50	21.30
		50–75	53.34
		75–100	34.28
		100–125	2.98
		> 125	1.25
2	Summer (April – June)	< 100	13.28
		100–125	43.22
		125–150	51.25
		150–175	4.35
		> 175	1.07
3	Pre-monsoon (July – September)	< 100	2.37
		100–125	10.85
		125–150	23.75
		150–175	43.87
		> 175	32.30
4	Monsoon (October – December)	< 50	19.27
		50–75	46.03
		75–100	40.04
		100–125	5.53
		> 125	2.27

TABLE 9.5
Extents of Distribution of the Gaseous and Particulate Pollutants (After Superimposition)

Sl No.	Pollutant	Concentration Range (μg/m³)	Extent (Km²)
1	Gaseous	Low	53.40
		Medium	39.48
		High	20.31
2	Particulate	Low	66.71
		Medium	34.35
		High	12.10

The concentration ranges of SO_2 that are taken for GIS study are 8.8–38.3 μg/m³, 6.5–125.3 μg/m³, 9.9–61.3 μg/m³, and 6.3–92.0 μg/m³ for the post-monsoon, summer, pre-monsoon, and monsoon seasons, respectively. The concentration ranges of NO_x that are taken for GIS study are 9.0–70.7 μg/m³, 9.0–129.7 μg/m³, 9.0–104.6 μg/m³, and 8.2–80.6 μg/m³ for the post-monsoon, summer, pre-monsoon, and monsoon seasons, respectively. Similarly for RSPM and TSP, the concentration ranges taken are

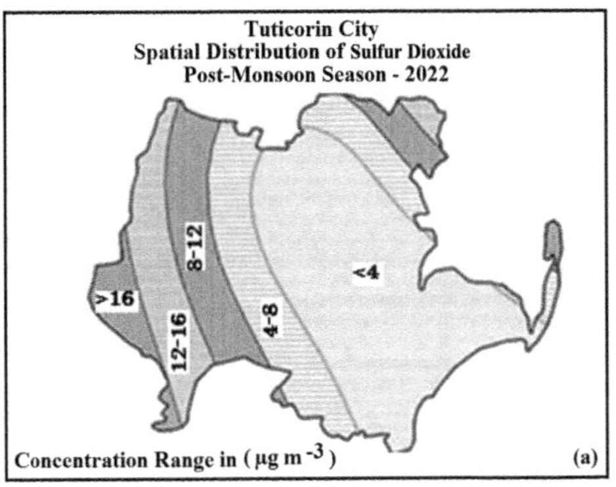

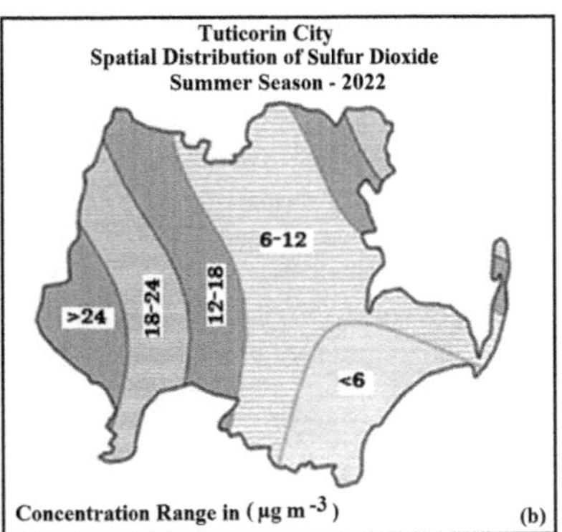

FIGURE 9.4 Spatial distribution of sulfur dioxide in post-monsoon season and summer season in Tuticorin city.

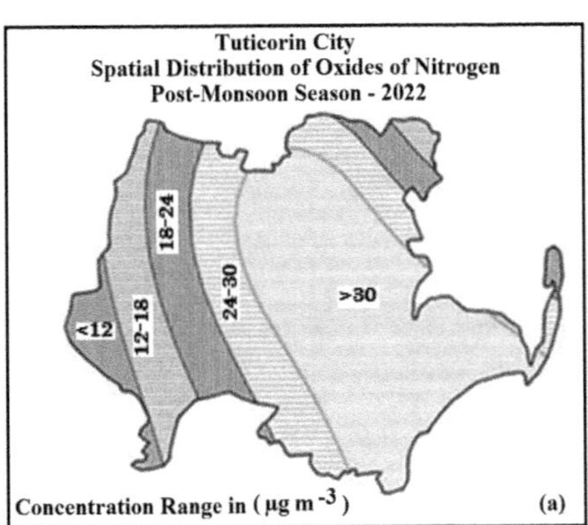

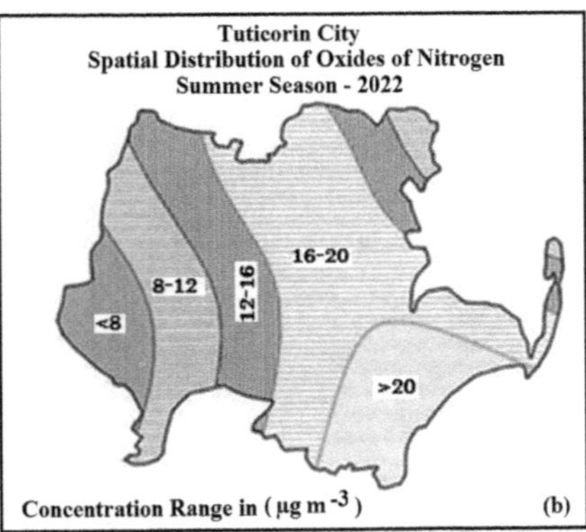

FIGURE 9.5 Spatial distribution of oxides of nitrogen in post-monsoon season and summer season in Tuticorin city.

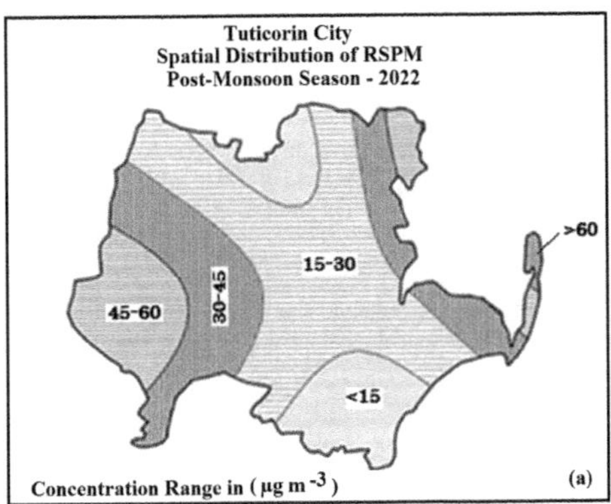

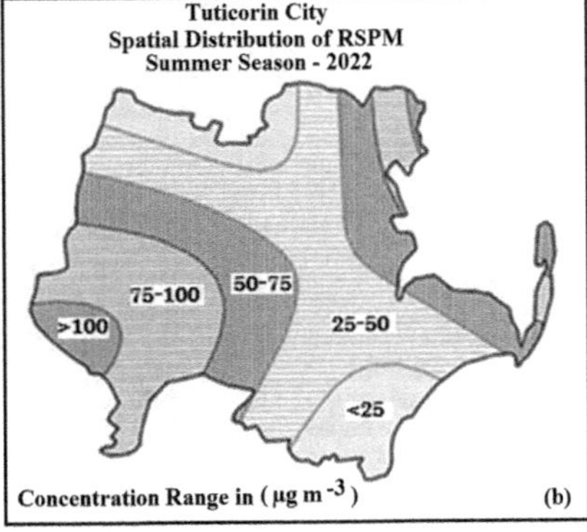

FIGURE 9.6 Spatial distribution of RSPM in post-monsoon season and summer season in Tuticorin city.

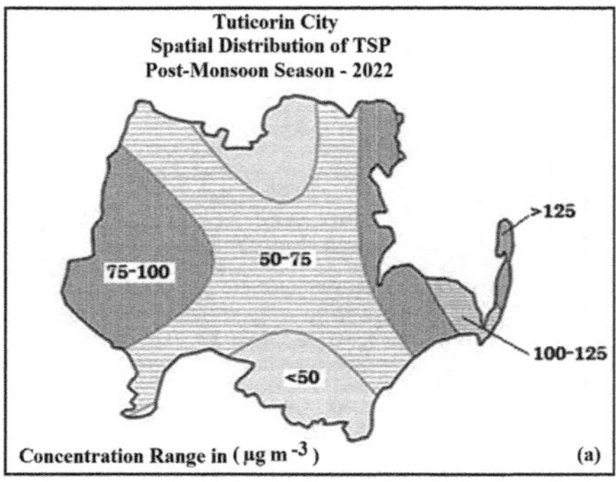

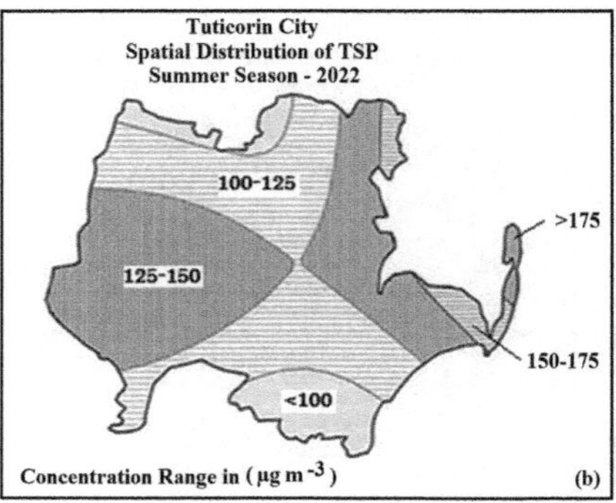

FIGURE 9.7 Spatial distribution of total suspended particulate matter (TSP) in post-monsoon season and summer season in Tuticorin city.

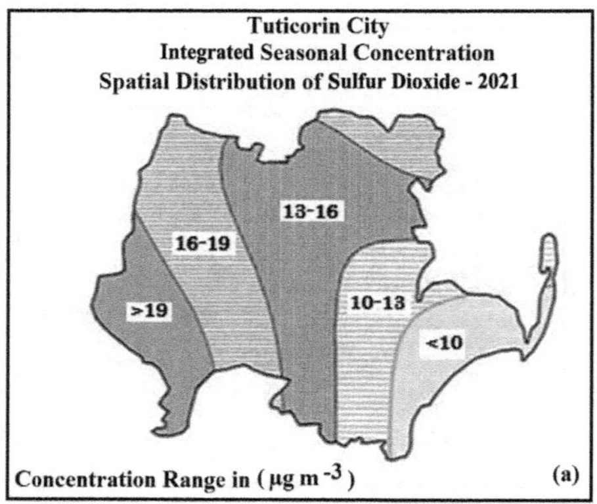

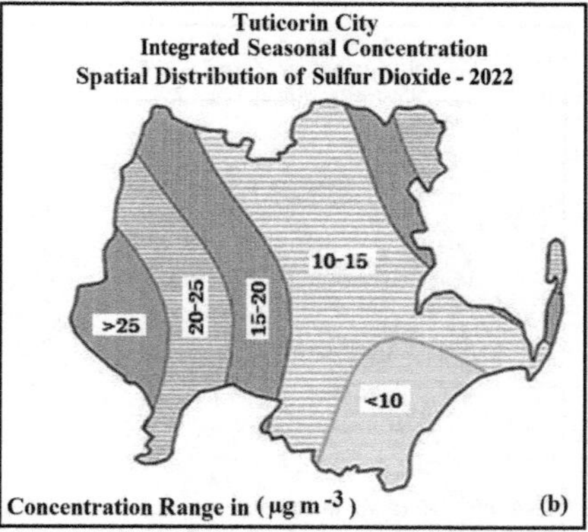

FIGURE 9.8 The integrated seasonal concentration of sulfur dioxide in Tuticorin city.

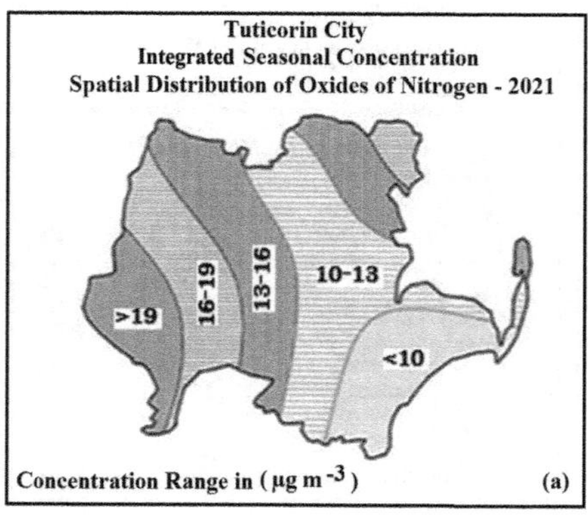

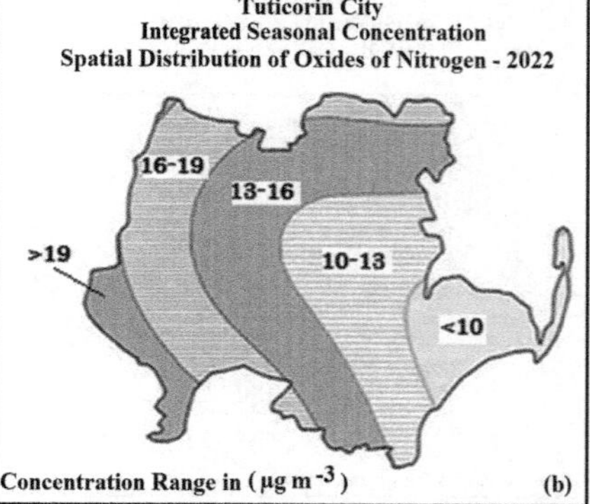

FIGURE 9.9 Integrated seasonal concentration of oxides of nitrogen in Tuticorin city.

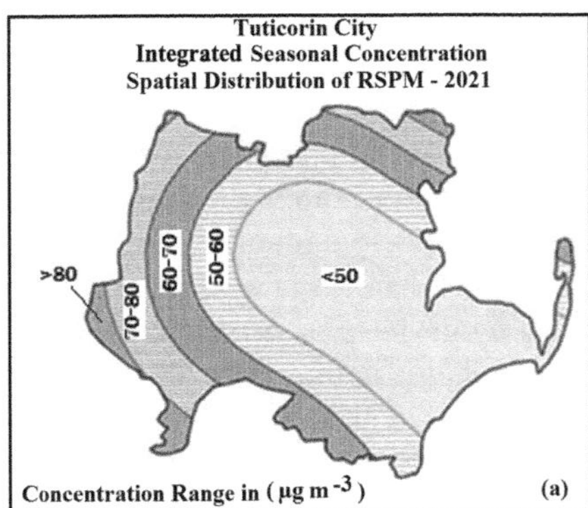

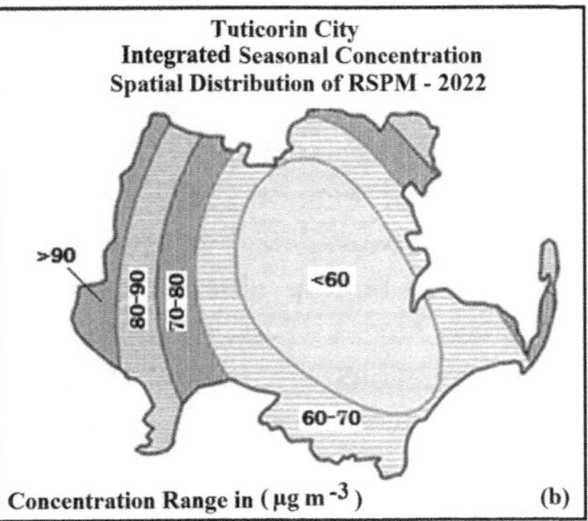

FIGURE 9.10 Integrated seasonal concentration of respirable suspended particulate matter (RSPM) in Tuticorin city.

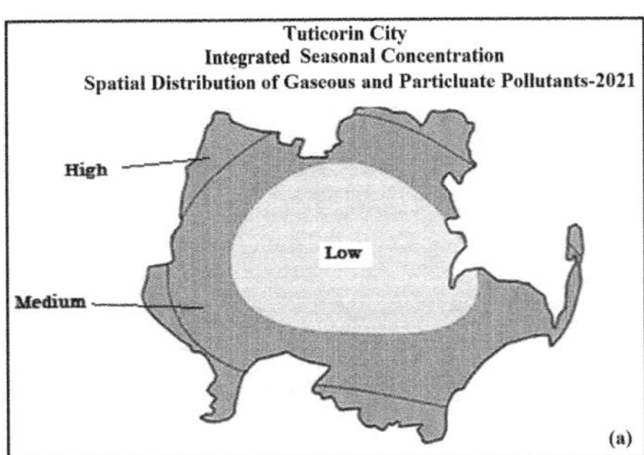

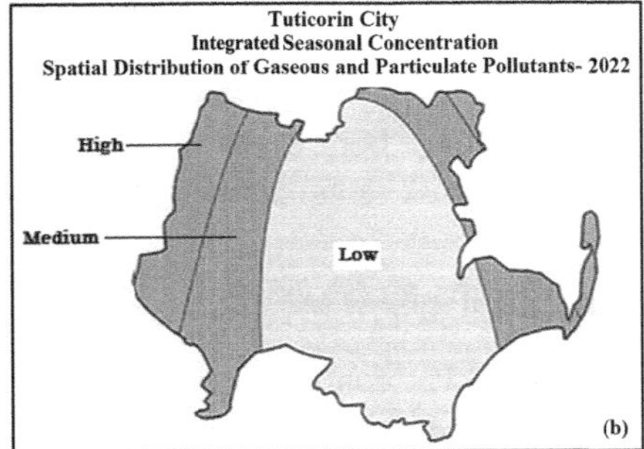

FIGURE 9.11 Integrated seasonal concentration of gaseous and particulate pollutants in Tuticorin city.

18–310 μg/m³ and 31–452 μg/m³, 25–423 μg/m³ and 42–548 μg/m³, 17–351 μg/m³ and 35–660 μg/m³, and 23–306 μg/m³ and 34–390 μg/m³ for the abovementioned seasons, respectively. The concentrations recorded in these seasons are divided into five ranges of equal intervals irrespective of their minimum and maximum values for the spatial display and subsequent interpretation of the distribution pattern of the pollutants.

In the year 2019, the data for the post-monsoon and summer seasons are not available for some of the study sites. Hence, the distribution pattern of the pollutants could not be generated for these seasons. In post-monsoon season, the SO_2 concentration range of 12–15 μg/m³ has the maximum extent of distribution, which is about 30% of the total area, and a range of less than 4 μg/m³ has the maximum distribution extent of about 45% of the total area in the years 2021 and 2022, respectively. In the same season, the NO_x concentration range of 12–15 μg/m³ has the maximum extent of distribution, which is around 30% of the total area, and a range of less than 12 μg/m³ has the maximum distribution extent of about 45% of the total area. For RSPM, the range of 15–30 μg/m³ has the maximum distribution extent, which is about 40% of the total area both in the years 2021 and 2022, and for TSP, the ranges of 30–45 μg/m³ and 50–75 μg/m³ have the maximum extent of distribution, which are about 35% and 47% of total area in the years 2021 and 2022 respectively.

In summer season, the SO_2 concentration range of 6–12 μg/m³ has the maximum extent of distribution, which is about 37% of the total area both in the years 2021 and 2022, respectively. The NO_x concentration range of 12–16 μg/m³ has the maximum extent of distribution, which is around 32% of the total area, in the same season. For RSPM, the ranges of 45–65 μg/m³ and 25–50 μg/m³ have the maximum extent of distribution, which are about 29% and 35% of total area in the years 2021 and 2022, respectively, and for TSP, the ranges of 25–50 μg/m³ and 125–150 μg/m³ have the maximum extent of distribution, which are about 39% and 45% of total area in the years 2021 and 2022, respectively.

In pre-monsoon season, the SO_2 concentration range of 12–18 μg/m³ covers the maximum extent of distribution both in the years 2019 and 2020, and the range 6–12 μg/m³ covered the maximum extent in the year 2022. The NO_x concentration ranges of 13–16 μg/m³, 15–18 μg/m³, and 5–10 μg/m³ have the maximum extent of distribution in the years 2019, 2021, and 2022, respectively, in the same season. For RSPM, the ranges of 45–65 μg/m³ both in the years 2021 and 2022 and 75–100 μg/m³ in the year 2022 occupy the maximum extent of distribution, and for TSP, the ranges of 100–125 μg/m³ both in the years 2021and 2022 and 150–175 μg/m³ in 2022 occupy the maximum extent of distribution.

9.4 DISCUSSION

Usually the PM contains solid and liquid particles such as dirt, soot, dust, and smoke. These sources may develop through different kinds of industries, manufacturing processes, and human activities against the nature. The industries such as copper industries, chemical industries, pharmacy industries, and cement factories are the potential sources considered in this analysis [28]. In general, in dry weather, the dust particles have more flying capability. Furthermore, in wet condition, the PMs have less flying capability due to the density variation. These parameters are correlated with the weather conditions [29].

The high concentration class occurs in and around Fisheries College and State Industries Promotion Corporation of Tamil Nadu (SIPCOT) area. So, critical care should be given immediately to pollution abatement in these areas so as to safeguard the people around these areas from adverse health effects of the pollutants [30]. Most parts of the city is experiencing with medium and low concentrations, and continuation of the pollution without mitigate measures may lead to high concentrations in future, which may adversely affect this urban environment.

These readily available and easily interpretable maps, by which any one can understand the degree of pollution and the extents of distribution of the pollutants, may be used for multiuser needs such as suitability analysis of the areas for developmental activities, policy- and decision-making, and urban planning. Suitable abatement measures may be framed and implemented for improving the air quality of the areas where there are high concentration distribution patterns of the pollutants [31].

The ear marked medium concentration areas have to be given proper concern to avoid further deterioration. Developmental activities, if necessary within the urban limits of the city, may be encouraged in the areas best suited for it, i.e., where the quality of ambient air is good. Keeping the 3M aspects such as monitoring, modeling, and management (using GIS tools) of the present thesis in holistic view and incorporating their own data in the scenario, various user agencies, such as transport department, town planning, pollution control board, voluntary organizations etc., collectively may develop an integrated package that may include suitable strategies for the management of the atmospheric carrying capacity, abatement strategies that may maintain healthy balance, and developmental strategies that may retain harmony with nature for protecting the health and both the personal and social wealth of the people living in the city [32].

The area of 25.74 Km^2 in which the SO_2 concentration range of 12–18 μg/m^3 is observed is the greatest extent identified among all the concentration ranges during monsoon season. The NO_x concentration range of 18–24 μg/m^3 covered the area of 47.13 Km^2 is the greatest identified extent during this season. For RSPM, 15–30 μg/m^3 in the years 2019 and 2020 and 50–75 μg/m^3 in the year 2022 are the concentration ranges that cover the maximum extent of distribution. As far as TSP is concerned, the ranges 25–50 μg/m^3, 50–75 μg/m^3, and 75–100 μg/m^3 occupy almost equal extent of area in the year 2019. But, less than 25 μg/m^3 have the maximum extent of distribution in the year 2020, and 50–75 μg/m^3 covers the maximum distribution extent in the year 2022. The above spike is subject to concern than other area, for example, in Nsukka, Nigeria, those gaseous constructions reported lesser than this [33].

As it is known, all the seasonal maps of both these pollutants of present interest are superimposed individually to understand the annual pollution status and are displayed. In the case of the pollutant SO_2, it is found that the concentration range 10–15 μg/m^3 of this pollutant in the ambient air is found to be in the first place in extent followed by 15–20 μg/m^3 and 20–25 μg/m^3 concentration ranges. In percentage wise, the 10–15 μg/m^3 concentrations are found to spread in 40.1%, whereas 15–20 μg/m^3 and 20–25 μg/m^3 concentrations are found to spread in 20.8% and 17.9% of the total extent, respectively. In the case of NO_x, the concentration range 13–16 μg/m^3 of this pollutant in the ambient air is found to be in the first place in extent followed by 10–13 μg/m^3 and 16–19 μg/m^3 concentration ranges. In percentage wise, the 13–16 μg/m^3 concentrations are found to spread in 32.5%, whereas 10–13 μg/m^3 and 16–19 μg/m^3 concentrations are found to spread in 24.3% and 23.5% of the total extent, respectively.

Likewise, in the case of RSPM, the concentration range <60 μg/m^3 of this pollutant in the ambient air is found to be in the first place in extent followed by 60–70 μg/m^3 and 70–80 μg/m^3 concentration ranges. In percentage wise, these concentrations are found to spread in 34.8%, 31.6%, and 14.4% of the total extent, respectively. As far as TSP is concerned, the concentration range <100 μg/m^3 of this pollutant in the ambient air is found to be in the first place in extent followed by 100–120 μg/m^3 and 120–140 μg/m^3 concentration ranges. In percentage wise, these concentrations are found to spread in 47.7%, 26.9%, and 16.5% of the total extent, respectively. In recent decades, the applied policies as well as the several lockdowns in pandemic period have minimized the amount of it [34].

It is observed from these maps that the sources of the high NO_x and SO_2 concentrations are found in the industrial zone and hence are mostly due to the industrial activities. It is obvious from the integrated map that the medium NO_x concentrations are widespread, particularly in the central parts of the city due to the operation of diesel driven vehicles. As far as particulate pollutants are concerned, as we move away from the center part of the city, the concentration increases. This might be due to the fact that the center part of the city is mostly covered by roads and buildings and have less industrial activities and hence less chance for having resuspended dust when compared to the other parts.

As it is known, the final integrated maps are obtained by overlapping the superimposed maps of both the gaseous pollutants and both the particulate pollutants for identifying the high, medium, and low concentrations areas of the study area. It is found from the map that low concentration areas are found to be in the first place in extent followed by medium and high concentration areas for both gaseous and particulate pollutants. In the case of gaseous pollutants, in percentage wise, the low concentrations are found to spread in 47.2% while the medium and high concentrations are found to spread in 34.9% and 17.9% of the total extent of the city, respectively. As far as particulate pollutants are concerned, the low concentrations are found to spread in 59.0% while the medium and high concentrations are found to spread in 30.3% and 10.7% of the total extent of the city, respectively. Medium and high concentration areas contribute 52.8% and 41% to the gaseous and particulate pollution, respectively. This suggests urgent measures to be taken in these particular areas.

9.5 CONCLUSIONS

It has been concluded from the study that suitable abatement measures need to be framed and implemented for improving the air quality of the areas where there are high concentration distribution patterns of the pollutants. The ear marked medium concentration areas have to be given proper concern to avoid further deterioration. Developmental activities, if necessary within the urban limits of the city, may be encouraged in the areas best suited for it, i.e., where the quality of ambient air is good. Keeping the aspects such as monitoring, modeling, and management (using GIS tools) in holistic view and incorporating their own data in the scenario, various user agencies, such as transport department, town planning, pollution control board, voluntary organizations etc., collectively may develop an integrated package that may include suitable strategies for the management of the atmospheric carrying capacity, abatement strategies that may maintain healthy balance, and developmental strategies that may retain harmony with nature for protecting the health and both the personal and social wealth of the people living in the city. Moreover, the prospective study needs to explore the different geostatistical methods (kriging and nongeostatistical such as Inverse distance weighting (IDW)) and will establish a significant comparative study between those applied methods for the spatial distribution of pollutants.

REFERENCES

1. Brown KW, Sarnat JA, Suh HH, Coull BA, Koutrakis P. Factors influencing relationships between personal and ambient concentrations of gaseous and particulate pollutants. *Sci Total Environ.* 2009;407: 3754–3765.
2. Sathishkumar RS, Sundaramanickam A, Mohanty AK, Sahu G, Ramesh T, Balachandar K, et al. Seasonal assessment of the trophic status in the coastal waters adjoining Tuticorin harbor in relation to water quality and plankton community in the Gulf of Mannar, India. *Oceanologia.* 2022;64: 749–768.
3. Rajaram R, Ganeshkumar A, Vinothkannan A. Health risk assessment and bioaccumulation of toxic metals in commercially important finfish and shellfish resources collected from Tuticorin coast of Gulf of Mannar, Southeastern India. *Mar Pollut Bull.* 2020;159: 111469.
4. Ariola V, D'alessandro A, Lucarelli F, Marcazzan G, Mazzei F, Nava S, et al. Elemental characterization of PM10, PM2.5 and PM1 in the town of Genoa (Italy). *Chemosphere.* 2006;62: 226–232.
5. Berkowicz R, Winther M, Ketzel M. Traffic pollution modelling and emission data. *Environ Model Softw.* 2006;21: 454–460.
6. Ji H, Shao M, Wang Q, others. Contribution of meteorological conditions to inter-annual variations in air quality during the past decade in Eastern China. *Aerosol Air Qual Res.* 2020;20: 2249–2259.
7. Jo W-K, Park J-H. Characteristics of roadside air pollution in Korean metropolitan city (Daegu) over last 5 to 6 years: temporal variations, standard exceedances, and dependence on meteorological conditions. *Chemosphere.* 2005;59: 1557–1573.
8. Harrison RM, Yin J. Particulate matter in the atmosphere: which particle properties are important for its effects on health? *Sci Total Environ.* 2000;249: 85–101.
9. Chuersuwan N, Nimrat S, Lekphet S, Kerdkumrai T. Levels and major sources of PM2.5 and PM10 in Bangkok Metropolitan Region. *Environ Int.* 2008;34: 671–677.
10. Das IK, Rajendrakumar P. Disease resistance in sorghum. *Biotic Stress Resistance in Millets.* Elsevier, Amsterdam, The Netherlands; 2016, pp. 23–67.
11. Fu JS, Jang CJ, Streets DG, Li Z, Kwok R, Park R, et al. MICS-Asia II: Modeling gaseous pollutants and evaluating an advanced modeling system over East Asia. *Atmos Environ.* 2008;42: 3571–3583.
12. Hong Y-M, Lee B-K, Park K-J, Kang M-H, Jung Y-R, Lee D-S, et al. Atmospheric nitrogen and sulfur containing compounds for three sites of South Korea. *Atmos Environ.* 2002;36: 3485–3494.
13. Dalvi M, Beig G, Patil U, Kaginalkar A, Sharma C, Mitra AP. A GIS based methodology for gridding of large-scale emission inventories: application to carbon-monoxide emissions over Indian region. *Atmos Environ.* 2006;40: 2995–3007.
14. Bedi JS, Dhaka P, Vijay D, Aulakh RS, Gill JPS, others. Assessment of air quality changes in the four metropolitan cities of India during COVID-19 pandemic lockdown. *Aerosol Air Qual Res.* 2020;20: 2062–2070.
15. Chang S-C, Lee C-T. Evaluation of the trend of air quality in Taipei, Taiwan from 1994 to 2003. *Environ Monit Assess.* 2007;127: 87–96.
16. Zahraee SM, Shiwakoti N, Stasinopoulos P. Application of geographical information system and agent-based modeling to estimate particle-gaseous pollutant emissions and transportation cost of woody biomass supply chain. *Appl Energy.* 2022;309: 118482.
17. Nonnemacher M, Jakobs H, Viehmann A, Vanberg I, Kessler C, Moebus S, et al. Spatio-temporal modelling of residential exposure to particulate matter and gaseous pollutants for the Heinz Nixdorf Recall Cohort. *Atmos Environ.* 2014;91: 15–23.
18. Sharma N, Bhandari K, Rao P, Shukla A. GIS applications in air pollution modeling. *Proceedings of the 6th International Conference 'Map India 2003.* 2003;2003: 28–31.

19. Zhang Y, Ding Z, Xiang Q, Wang W, Huang L, Mao F. Short-term effects of ambient PM1 and PM2.5 air pollution on hospital admission for respiratory diseases: case-crossover evidence from Shenzhen, China. *Int J Hyg Environ Health.* 2020;224: 113418.
20. Guo L, Chen Y, Mi B, Dang S, Zhao D, Liu R, et al. Ambient air pollution and adverse birth outcomes: a systematic review and meta-analysis. *J Zhejiang Univ Sci B.* 2019;20: 238.
21. Lupia F. Erdas Imagine 9.2: An Overview of the Main Features and Tools. Technical Report, 2012. DOI: 10.13140/2.1.3689.9525.
22. Navinya CD, Vinoj V, Pandey SK, others. Evaluation of PM2.5 surface concentrations simulated by NASA's MERRA version 2 aerosol reanalysis over India and its relation to the air quality index. *Aerosol Air Qual Res.* 2020;20: 1329–1339.
23. Orellano P, Reynoso J, Quaranta N, Bardach A, Ciapponi A. Short-term exposure to particulate matter (PM10 and PM2.5), nitrogen dioxide (NO_2), and ozone (O_3) and all-cause and cause-specific mortality: systematic review and meta-analysis. *Environ Int.* 2020;142: 105876.
24. Punsompong P, Pani SK, Wang S-H, Pham TTB. Assessment of biomass-burning types and transport over Thailand and the associated health risks. *Atmos Environ.* 2021;247: 118176.
25. Shahbazi H, Abolmaali AM, Alizadeh H, Salavati H, Zokaei H, Zandavi R, et al. Development of high-resolution emission inventory to study the relative contribution of a local power plant to criteria air pollutants and Greenhouse gases. *Urban Clim.* 2021;38: 100897.
26. Zhang Y, Fang J, Mao F, Ding Z, Xiang Q, Wang W. Age-and season-specific effects of ambient particles (PM1, PM2.5, and PM10) on daily emergency department visits among two Chinese metropolitan populations. *Chemosphere.* 2020;246: 125723.
27. Yarragunta Y, Srivastava S, Mitra D, Chandola HC. Source apportionment of carbon monoxide over India: a quantitative analysis using MOZART-4. *Environ Sci Pollut Res.* 2021;28: 8722–8742.
28. Tella A, Balogun A-L. GIS-based air quality modelling: spatial prediction of PM10 for Selangor State, Malaysia using machine learning algorithms. *Environ Sci Pollut Res.* 2021;29: 1–17.
29. Deshmukh DK, Deb MK, Mkoma SL. Size distribution and seasonal variation of size-segregated particulate matter in the ambient air of Raipur city, India. *Air Qual Atmos Heal.* 2013;6: 259–276.
30. Selvam S, Antony Ravindran A, Venkatramanan S. Assessment of heavy metal and bacterial pollution in coastal aquifers from SIPCOT industrial zones, Gulf of Mannar, South Coast of Tamil Nadu, India. *Appl Water Sci.* 2017;7: 897–913.
31. Gulia S, Shukla N, Padhi L, Bosu P, Goyal SK, Kumar R. Evolution of air pollution management policies and research gap assessment in India. *Environ Challenges.* 2021;6: 100431.
32. Jumaah HJ, Ameen MH, Kalantar B, Rizeei HM, Jumaah SJ. Air quality index prediction using IDW geostatistical technique and OLS-based GIS technique in Kuala Lumpur, Malaysia. *Geomat Nat Hazards Risk.* 2019;10: 2185–2199.
33. Agbo KE, Walgraeve C, Eze JI, Ugwoke PE, Ukoha PO, Van Langenhove H. Household indoor concentration levels of NO_2, SO_2 and O_3 in Nsukka, Nigeria. *Atmos Environ.* 2021;244: 117978.
34. Sivaramasundaram K, Muthusubramanian P. A preliminary assessment of PM10 and TSP concentrations in Tuticorin, India. *Air Qual Atmos Heal.* 2010;3: 95–102.

10 Technological Innovation and Its Impact on Carbon Emissions

Manufacturing Industries

K. Harisivasri Phanindra and S.P. Sivapirakasam

10.1 INTRODUCTION

Carbon emissions refer to the release of carbon-containing gases into the atmosphere, mainly carbon dioxide (CO_2), and also nitrous oxide (N_2O) and methane (CH_4), among others [1]. These emissions stem from various human activities, with the burning of fossil fuels for transportation, production, and energy creation being the primary contributors [2]. Manufacturing industries are substantial contributors to carbon emissions due to the nature of their operations and reliance on energy-intensive processes [3]. The main sources of carbon emissions in manufacturing include the process of burning fossil fuels to produce energy and heat production [4]. This encompasses the use of coal, oil, and natural gas in boilers, furnaces, and other industrial equipment to power manufacturing processes, such as smelting, refining, and chemical synthesis. Additionally, the production of raw materials, such as steel, cement, and chemicals, involves chemical reactions that release CO_2 as a byproduct [3]. Furthermore, the transportation of raw materials and finished goods, often reliant on diesel-powered trucks and ships, also contributes to carbon emissions. Industrial activities like waste incineration and wastewater treatment can release CH_4 and N_2O, potent greenhouse gases (GHGs), further adding to the carbon footprint of manufacturing operations.

Carbon emissions have profound effects on human health and the environment, contributing significantly to climate change and exacerbating a range of environmental challenges. As GHGs, carbon emissions cause the Earth's atmosphere to retain heat, which raises global temperatures, shifts in weather conditions, and more frequent, harsh, and unusual weather conditions [5]. Ecosystems are disrupted by these changes, which results in habitat loss, changed migratory patterns, and a decline in biodiversity. Beyond environmental consequences, carbon emissions also pose significant risks to human health. Fossil fuel combustion results in the emission of nitrogen oxides (NO_X), particulate matter (PM), and volatile organic compounds (VOCs), among other air pollutants that are harmful to the quality of the air. Exposure to these pollutants is associated with respiratory diseases, cardiovascular conditions, and premature mortality. Overall, addressing carbon emissions is essential not only

DOI: 10.1201/9781003452072-10

for reducing the effects of climate change but also for safeguarding human health and preserving ecosystems [6].

In order to reach sustainability in terms of the environment, economy, and society, carbon emissions must be reduced. Fundamentally, sustainability entails meeting present needs without compromising the ability of future generations to meet their own requirements. Moreover, addressing carbon emissions is essential for fostering economic sustainability. Making a shift to renewable energy sources and putting energy-saving measures in place not only reduces reliance on finite and environmentally harmful resources but also stimulates innovation, creates green jobs, and enhances economic resilience [7]. Industries that embrace low-carbon practices often experience cost savings through reduced energy consumption, waste reduction, and regulatory compliance, while also attracting investment and remaining competitive in a rapidly evolving market shaped by sustainability concerns. In essence, reducing carbon emissions is not only an environmental imperative but a fundamental prerequisite for achieving sustainability across all sectors of society. By embracing cleaner, more efficient technologies, transitioning to renewable energy sources, and fostering collaboration among governments, industries and civil society help create a more inclusive and sustainable future for coming generations.

The main objective of this chapter is to understand the carbon emissions landscape within manufacturing industries, particularly in sectors like cement, aluminum, paper and pulp, and iron and steel production. It aims to explore how technological innovations, drawn from scholarly literature, can help reduce carbon emissions from manufacturing processes.

10.2 OVERVIEW OF CEMENT INDUSTRY

Cement making in the world reached 4.1 billion tons in 2018, according to data from the United States Geological Survey (USGS) as shown in Figure 10.1. China remains the largest producer, accounting for 57.8% of the overall production, followed by India with 7.1%. Additional noteworthy producers consist of Brazil (2.2%), United States (2.2%), Turkey (2%), Vietnam (2%), Indonesia (1.6%), Japan (1.4%), Iran (1.3%), Russia (1.3%), Republic of Korea (1.1%), Saudi Arabia (1.1%), and various other countries contributing 18.5% [8].

Cement stands as the fundamental cornerstone in civil engineering, serving as the most fundamental and widely utilized building material. With the rapid pace of urbanization, its demand has surged dramatically. However, the cement industry stands as a significant part of carbon emissions. In 2006 alone, this sector was responsible for approximately 1.8 gigatons of CO_2 emissions, constituting around 7% of the total human-caused CO_2 emissions globally [9].

Ishak et al. [10] have reviewed CO_2 emissions across various phases of cement making, encompassing processing of raw ingredients, manufacture of clinker, kiln fuel combustion, and final cement production, and revealed that the manufacturing of clinker accounted for 90% of CO_2 emissions from cement plants, rest 10% coming from the stages of cement production that involve finishing and preparing raw materials [10]. A survey of different CO_2 emission reduction options was conducted by Ishak et al. [10]. These strategies included enhancing energy efficiency, recovering waste

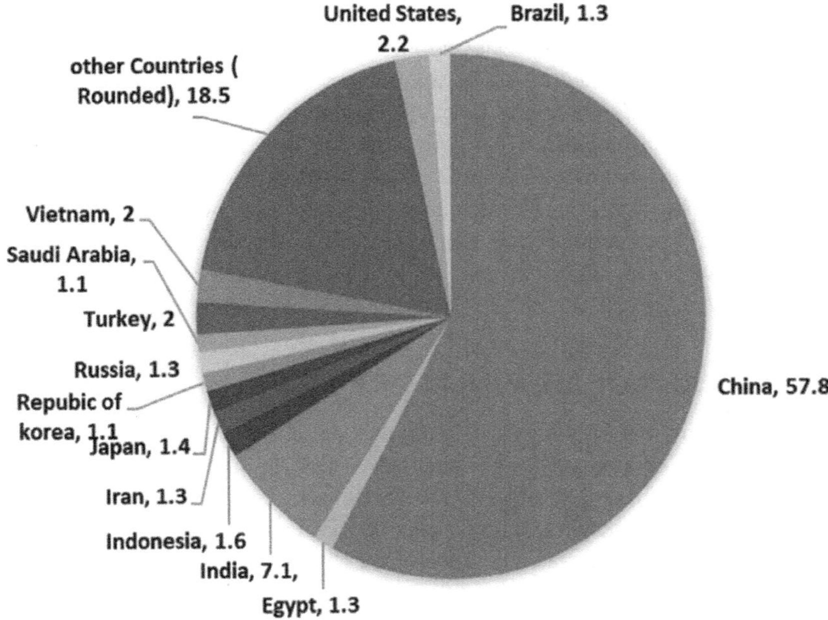

FIGURE 10.1 A graphic representation of the global percentage share of cement production and the major contributing nations [8].

heat, utilizing renewable energy sources instead of fossil fuels, making low-carbon cement, and carbon capture and storage (CCS) [10]. Furthermore, the use of additional cementitious materials, such as fly ash, silica fume, sewage sludge, copper slag, and ground-granulated blast furnace slag, to lower carbon emissions [11].

China stands as the foremost producer and emitter of CO_2 within the global cement industry. Cement production in China accounts for a substantial 14.8% of the nation's CO_2 emissions, making it a pivotal sector in China's pursuit of its national goal of reducing carbon emissions by 40%–45% [12]. Drawing upon statistics from 15 cement production units, Gao et al. [9] demonstrated that the substitution of carbonate-containing materials with non-carbonate alternatives and adjustments to the clinker proportion were key avenues for reducing CO_2 content in raw materials and process emissions. For instance, the adoption of sulphoaluminate cement manufacturing in new-generation cement plants can yield reductions till 35% in CO_2 emissions per unit mass of cement produced, compared to conventional Portland cement manufacturing [9].

Gupta et al. [8] discussed the importance of sustainability in cement production, mainly through the growth of Portland-limestone cement (PLC). Gupta et al. [8] highlighted how the incorporation of limestone into cement can reduce the clinker factor, leading to significant fuel and energy savings and conservation of natural resources. Additionally, Gupta outlined the benefits of PLC, providing lower energy requirements for manufacturing, better workability, and a decrease in GHG emissions. By promoting the adoption of PLC, the cement industry can enhance its environmental sustainability and contribute to global efforts in climate action and achieving Sustainable Development Goals [8]. Mishra et al. [13] further summarized the difficulties faced by the cement sector, emphasizing the urgent need for adopting sustainable practices and technological innovations to mitigate environmental pollution and carbon emissions. Mishra et al. [13] underscored the importance of comprehensive research to assess the economic viability and environmental benefits of alternative fuels and materials in cement production. These insights highlight the complexities and opportunities for transitioning the cement industry toward a low-carbon future while addressing the growing demand for cement globally [13].

10.3 OVERVIEW OF ALUMINUM INDUSTRY

Globally, the aluminum industry contributes approximately 1% to total carbon emissions. China has experienced a significant surge in primary aluminum production over the past decade, reaching nearly 22 million metric tons in 2013, accounting for approximately 41% of the world's total primary aluminum output.

Brough et al. [14] conducted a comprehensive literature survey on the aluminum industry, providing valuable insights into its technological advancements, environmental effects, and potential for waste heat recovery. Figure 10.2 shows the material flow of the aluminum industry. The survey delved into the entire production process of aluminum, elucidating the state-of-the-art technologies employed at each stage, with a particular focus on casting technologies and secondary recycling [14]. Brough et al.'s [14] analysis highlighted the increasing proportion of recycled aluminum in the industry, underscoring its energy efficiency compared to primary production methods. Moreover, the survey explored future developments within the sector, including the promising prospects of inert anode technology in reducing CO_2 emissions [14]. In addition to technological advancements, Brough et al. [14] examined the environmental impacts associated with aluminum production, shedding light on gaseous emissions and solid residue by-products. The survey emphasized the industry's high energy intensity and the potential for waste heat recovery technologies

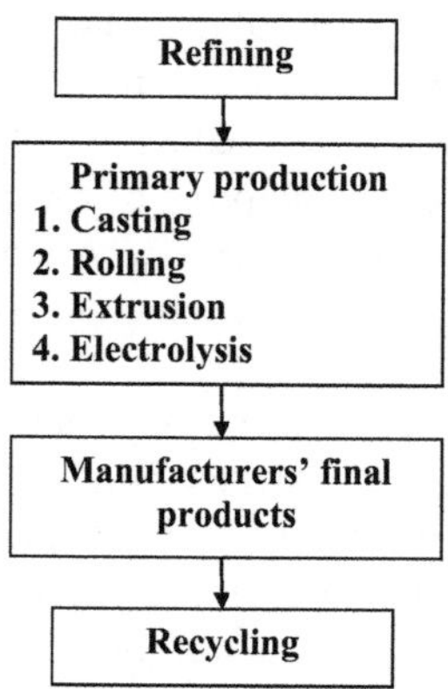

FIGURE 10.2 Typical process of aluminum industry.

to mitigate energy consumption and environmental impact. By scrutinizing various methods for reducing energy consumption and assessing the feasibility of waste heat recovery technologies, Brough provided valuable insights into sustainable practices within the aluminum industry [14].

Brough et al.'s [14] literature survey also underscore the evolving landscape of aluminum production, from traditional methods to innovative, energy-efficient processes. The survey highlights the imperative for the industry to embrace sustainable practices and leverage technological innovations to address its environmental footprint effectively [14]. With the emergence of inert anode technology and the growing prominence of secondary metal production, the aluminum industry is poised for transformative change toward a more sustainable future. Through the integration of waste heat recovery technologies and ongoing research efforts, the industry can enhance its energy efficiency and cut GHG emissions to create a more sustainable and environmentally friendly aluminum sector [14].

10.4 OVERVIEW OF THE PAPER AND PULP INDUSTRY

The pulp and paper industry plays a crucial role in global GHG emissions, contributing approximately 3.2% of anthropogenic emissions from 1961 to 2019. Figure 10.3 illustrates the global GHG emissions from the paper industry from 1961 to 2019, highlighting emissions during various stages such as gathering raw materials, pulping, creating and printing paper, using, and disposing of waste [15].

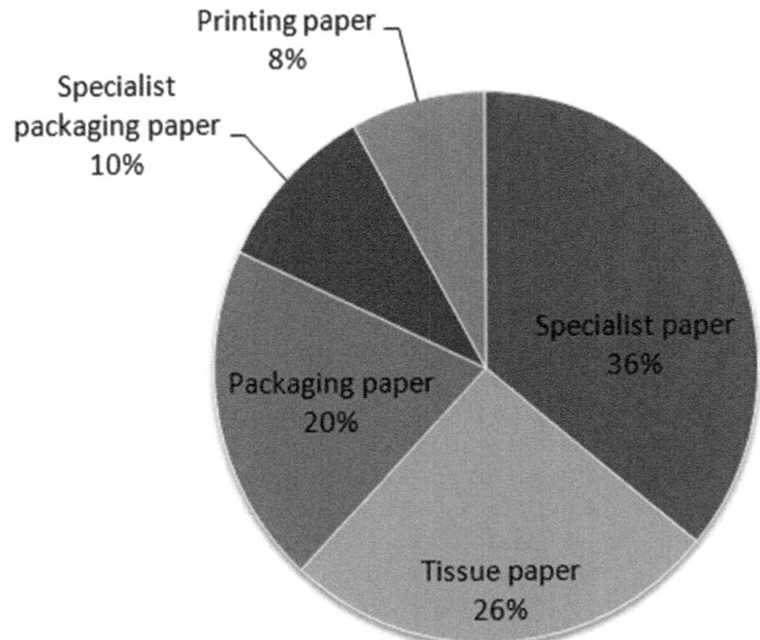

FIGURE 10.3 Global GHG emissions from the paper industry from 1961 to 2019.

Reaching net-zero emissions by 2050 requires tailored strategies for each country due to the industry's diverse nature across different nations. Through a meticulous bottom-up assessment covering 30 major countries, it becomes evident that historical emission trends and structural disparities vary significantly. Energy efficiency improvements and the decarbonization of energy systems emerge as critical priorities, particularly for developing countries. Sustainable forest management is of paramount importance for tropical nations abundant in forests, whereas others should focus on CH_4 capture and recycling enhancements [15].

A systems approach underscores opportunities beyond production stages, emphasizing the importance of carbon stocks in waste management. Developed countries demonstrate significant emission reduction potential through sustainable forestry practices, advanced technology adoption, and efficient waste recovery systems [15]. Priority should be given to energy-related measures, leveraging the best available techniques and fostering international cooperation. Sustainable raw material collection and improved waste management practices are essential components of net-zero strategies. Commitments to zero deforestation are crucial for forest-rich regions, ensuring the sustainable supply of raw materials [15].

Efficient waste sorting and disposal systems can yield notable negative emissions, complemented by diverse recycling strategies. It is imperative to incorporate the emissions data from various stages into the net-zero strategies. Incorporating S4 stage emissions benefits into carbon-trading systems fosters sustainable landfill practices and energy recovery, further contributing to the industry's overall sustainability goals. Overall, country-specific net-zero strategies are pivotal for the pulp and paper industry to mitigate its environmental impact effectively and transition toward a sustainable future. By adopting tailored measures and leveraging advancements in technology and sustainable practices, countries can work toward achieving net-zero emissions while ensuring the continued growth and development of the pulp and paper industry [15]

10.5 OVERVIEW OF IRON AND STEEL PRODUCTION

Iron and steel production stands as a significant contributor to anthropogenic CO_2 emissions, posing challenges for global climate targets aimed at limiting the increase in mean temperature to a range of 2.4°C–3.2°C [16].

The steel industry is under pressure to curb its CO_2 emissions due to its significant carbon consumption. This has led to a global trend of transitioning iron and steel enterprises, with countries implementing tailored plans and reform measures suited to their specific contexts [16]. Table 10.1 summarizes a brief overview of the CO_2 reduction approaches. Efforts are primarily focused on two fronts: developing reducing agents with minimal carbon content and devising environmentally friendly methods for disposing the by-products from blast furnace operations.

Reducing the carbon content of ironmaking hinges on optimizing the percentage of reducing agents in low-blast furnaces, with minimizing coke usage as a top research priority. Maximizing the utilization of carbon chemical energy through the separation and recycling of CO_2 in furnace top gas shows promise for enhancing reducing agent efficiency. Additionally, exploring alternate energy sources like

TABLE 10.1
A Brief Overview of the CO_2 Reduction Approaches

Country/Continent	Projects	Target	Technologies
China	Hydrogen metallurgy, Hismelt and COREX	Achieving carbon neutrality by 2060	Elimination of the coking process and enhancement of gas utilization efficiency
The United States	Hydrogen flash smelting and Molten oxide electrolysis	Enhancing environmental sustainability and energy efficiency	Integration of CO_2 capture and storage technologies
Europe	ULCOS ULCORED, ULCOLYSIS, and ULCOWIN	Targeting a 50% or greater reduction in CO_2 emissions	Implementation of Vacuum pressure swing adsorption (VPSA) technology for CO_2 capture and conversion to methanol; elimination of coking and sintering in the melting reduction process; utilization of natural gas or coal gasification as reducing agents
Germany	Carbon2Chem and SALCOS	–	Conversion of CO_2 within exhaust gases
Austria	H2FUTURE	Production of CO_2-free hydrogen for various sectors	Electrolysis of water to produce hydrogen with a focus on its widespread availability
Sweden	HYBRIT	Achieving zero emissions within 10–25 years	As a reducing agent, hydrogen is used to reduce sponge iron directly
Japan	COURSE50	Reduction of carbon emissions by 30%	Reduction of carbon emissions by 10% through hydrogen utilization and 20% via separation and recovery of CO_2 from blast furnace gas
Korea	FINEX	Decreasing CO_2 emissions by 45% with CCS	Utilization of waste gas and by-products to generate hydrogen, leveraging marine biological waste for CO_2 sequestration, CO_2 capture using ammonia, waste heat reuse, and implementing low-carbon methods in sintered ore production

biomass fuel and waste fuel, alongside electric heating, offers avenues to reduce dependency on coke in the ironmaking process [16]

Efficient ironmaking processes rely on the decrease of iron oxides and the separation of slag and iron through slagging. While the blast furnace remains highly efficient, there is growing interest in exploring low-carbon routes for ironmaking, prompting ongoing research and development efforts. As a global steel production leader, China bears a significant responsibility in reducing emissions and advancing energy-saving processes. Investing in progressive, low-carbon technologies such as CCS, recycling of waste heat energy at high temperatures, and alternative reducing agents like hydrogen or biomass is crucial [16,17].

Despite available technologies to replace coke with pulverized coal, widespread adoption remains limited. Phasing out both coke and pulverized coal in favor of renewable energy sources is essential for long-term sustainability. Moreover, efficient extraction methods for carbon-containing by-products are crucial for reducing GHGs [16]. While these advancements signify progress toward a low-carbon emissions environment, continued research efforts are essential to overcome existing challenges for a sustainable future.

10.6 SUMMARY

This chapter delves into the intricate relationship between technological innovation and carbon emissions in manufacturing industries. It highlights the significant role of manufacturing in global carbon emissions due to energy-intensive processes and explores the environmental and health impacts of these emissions. It discusses strategies such as energy efficiency improvements, waste heat recovery, renewable energy adoption, and CCS. Moreover, this chapter emphasizes the importance of transitioning toward sustainable practices to achieve environmental, economic, and social sustainability goals.

REFERENCES

1. Paolini, V., Petracchini, F., Segreto, M., Tomassetti, L., Naja, N. and Cecinato, A., 2018. Environmental impact of biogas: A short review of current knowledge. *Journal of Environmental Science and Health, Part A*, 53(10), pp.899–906.
2. Oyewole, K.A., Okedere, O.B., Rabiu, K.O., Alawode, K.O. and Oyelami, S., 2023. Carbon dioxide emission, mitigation and storage technologies pathways. *Sustainable Environment*, 9(1), p.2188760.
3. Griffin, P.W., Hammond, G.P. and Norman, J.B., 2018. Industrial energy use and carbon emissions reduction in the chemicals sector: A UK perspective. *Applied Energy*, 227, pp.587–602.
4. Hu, P., Li, Y., Zhang, X., Guo, Z. and Zhang, P., 2018. CO_2 emission from container glass in China, and emission reduction strategy analysis. *Carbon Management*, 9(3), pp.303–310.
5. Gahlawat, I.N. and Lakra, P., 2020. Global climate change and its effects. *Integrated Journal of Social Sciences*, 7(1), pp.14–23.
6. Xue, X., Zhang, Q., Cai, X. and Ponkratov, V.V., 2023. Multi-criteria decision analysis for evaluating the effectiveness of alternative energy sources in China. *Sustainability*, 15(10), p.8142.

7. Jie, H., Zaman, S., Uz Zaman, Q., Shah, A.H. and Lou, J., 2023. A pathway to a sustainable future: Investigating the contribution of technological innovations, clean energy, and women's empowerment in mitigating global environmental challenges. *Journal of Cleaner Production*, 421, p.138499.
8. Gupta, S., Mohapatra, B.N. and Bansal, M., 2020. A review on development of Portland limestone cement: A step towards low carbon economy for Indian cement industry. *Current Research in Green and Sustainable Chemistry*, 3, p.100019.
9. Gao, T., Shen, L., Shen, M., Chen, F., Liu, L. and Gao, L., 2015. Analysis on differences of carbon dioxide emission from cement production and their major determinants. *Journal of Cleaner Production*, 103, pp.160–170.
10. Ishak, S.A. and Hashim, H., 2015. Low carbon measures for cement plant – A review. *Journal of Cleaner Production*, 103, pp.260–274.
11. Crossin, E., 2015. The greenhouse gas implications of using ground granulated blast furnace slag as a cement substitute. *Journal of Cleaner Production*, 95, pp.101–108.
12. Chen, W., Hong, J. and Xu, C., 2015. Pollutants generated by cement production in China, their impacts, and the potential for environmental improvement. *Journal of Cleaner Production*, 103, pp.61–69.
13. Mishra, U.C., Sarsaiya, S. and Gupta, A., 2022. A systematic review on the impact of cement industries on the natural environment. *Environmental Science and Pollution Research*, 29(13), pp.18440–18451.
14. Brough, D. and Jouhara, H., 2020. The aluminium industry: A review on state-of-the-art technologies, environmental impacts and possibilities for waste heat recovery. *International Journal of Thermofluids*, 1, p.100007.
15. Del Rio, D.D.F., Sovacool, B.K., Griffiths, S., Bazilian, M., Kim, J., Foley, A.M. and Rooney, D., 2022. Decarbonizing the pulp and paper industry: A critical and systematic review of sociotechnical developments and policy options. *Renewable and Sustainable Energy Reviews*, 167, p.112706.
16. Zhang, X., Jiao, K., Zhang, J. and Guo, Z., 2021. A review on low carbon emissions projects of steel industry in the world. *Journal of Cleaner Production*, 306, p.127259.
17. Zhang, W., Li, H., Chen, B., Li, Q., Hou, X. and Zhang, H., 2015. CO_2 emission and mitigation potential estimations of China's primary aluminum industry. *Journal of Cleaner Production*, 103, pp.863–872.

11 Smart Grid Technologies in Plant Operations for Waste Reduction

R. Vivek, S. Ganeshkumar, P. Arunachalam, and R. Rameshbabu

11.1 OVERVIEW OF SMART GRID TECHNOLOGIES

Smart grid technology is ushering in a new era of enhanced sustainability, reliability, and efficiency in electrical infrastructure. In essence, a smart grid is an intelligent, networked system that optimizes electricity generation, distribution, and usage via the use of advanced sensors, digital technology, and communication networks. Utilities can accurately track patterns in energy use and modify their services in response to changing demand by using real-time data analytics. Smart grids facilitate the smooth integration of renewable energy sources, such as solar and wind power, into the system, therefore promoting a more sustainable and cleaner mix of energy sources [1–5]. Automation and control technologies help the grid become more resilient by cutting down on downtime, optimizing workflows, and enabling fast corrections of mistakes or disruptions. All things considered, smart grid technologies pave the way for an energy infrastructure that is more ecologically conscious, efficient, flexible, and able to meet the ever-changing needs of the contemporary world. The purpose of grid technologies is to use the aggregate processing capacity of several PCs and servers connected to a network. These distributed resources might be dispersed throughout many businesses, geographical areas, or even continents. Grid computing enables the sharing of computer resources, such as memory, storage, processing power, and specialized hardware. Sharing this information can improve the efficiency with which resources are used, save money, and improve the performance of computationally intensive tasks. Virtual groups are more likely to form when resources from several organizations—such as universities, research institutions, or businesses—are merged to achieve common goals. These online businesses are able to allocate and manage resources in real time based on demand and availability [6].

11.2 LITERATURE REVIEW

The expanding need for electrical energy, which presents issues in production and distribution, is covered in this portion of the chapter. It presents the idea of "smart grids," outlining their features and implications for the electricity distribution sector.

DOI: 10.1201/9781003452072-11

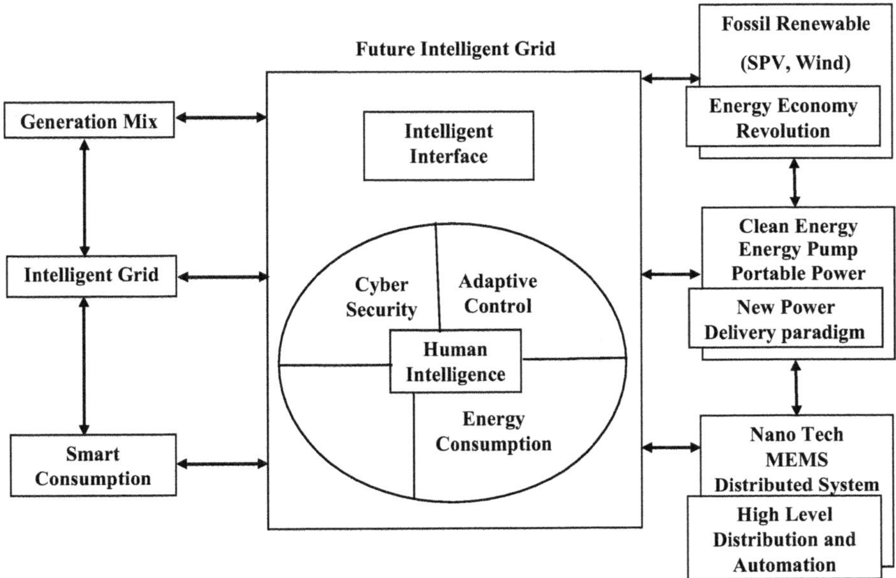

FIGURE 11.1 Methodology of smart grid technology and its applications [8].

This chapter highlights how these technologies improve the distribution system's sustainability, reliability, and efficiency and their potential for further development [7].

An in-depth analysis of smart grid technology and applications is covered in a study conducted by Bayindir in 2016 (*Renewable Energy*), with a focus on the enhanced capabilities of self-healing, flexibility, and sustainability. It covers a range of supporting technologies, including cloud computing, metering, connectivity, and apps used in smart grid systems [8]. Technologies including enhanced metering infrastructure, sophisticated electronic devices for communication, and sensors/vehicle-to-grid systems are examined. In addition to highlighting uses such as feeding automation and house automation, the report envisions financial gains and the creation of standards that work together. In addition to providing direction for upcoming development and coordination initiatives, it is an invaluable tool for researchers, engineers, and operators navigating the switch from conventional to smart grid systems. The graphical abstract of the same is shown in Figure 11.1.

Global problems surrounding energy sustainability and the environment owing to climate change and increasing energy consumption are reviewed in Tuballa et al.'s [9] study of the development of smart grid technology (*Renewable and Sustainable Energy Reviews*). It emphasizes distributed electricity production and microgrids as examples of sustainable technologies that are made possible by the smart grid. The relevance of research, policy support, and public awareness is emphasized in the report [7,9,10–17]. It gives a global overview of the characteristics, technologies, research projects, difficulties, and deployment strategies related to smart grids. Smart grid projects are viewed as cooperative rather than competing endeavors across various areas. This chapter highlights how adopting a smart grid may be facilitated by certain energy regulations and how smart grid technologies are always changing to better adapt to energy demand. Tuballa et al.'s [9] assessment of the development of smart grid technology (*Renewable and Sustainable Energy Reviews*) reviews global issues around energy sustainability and the environment due to climate change and growing energy usage. Microgrids and dispersed power generation are highlighted as instances of sustainable technology enabled by the smart grid. This chapter emphasizes the need of public awareness, policy support, and research. It provides a broad overview of the traits, advancements in technology, ongoing studies, challenges, and approaches to implementation of smart grids. Smart grid initiatives are seen as collaborative efforts rather than rival initiatives in different domains. This chapter describes how some energy rules may make it easier to establish a smart grid and how smart grid technologies are always changing to better adapt to energy demand. M. Fadaeenejad and colleagues' 2014 article, "*The Present and Future of Smart Power Grid in Developing Countries*" (*Renewable and Sustainable Energy Reviews*), explores this concept. Recent years have seen a considerable increase in interest in smart grid technology worldwide, indicating great prospects for future study. While poor nations are progressively looking into pilot projects and research efforts, developed nations confront obstacles in both research and execution. Based on their initiatives related to smart grids, emerging nations are divided as pioneers and others in this article. Significant progress has been made by nations like China, India, and Brazil, whose efforts are equivalent to those of wealthy nations like the United States. Their innovations provide other developing countries with benchmarks and ideas for more smart grid developments. These pioneering nations' achievements are a useful point of reference advice for other underdeveloped countries starting their smart grid projects. Emerging economies may learn about best practices, possible obstacles, and possibilities for smart grid adoption by examining the experiences and tactics of developed nations [9]. This information may help with decision-making and direct the creation of customized smart grid systems that fit the unique requirements and conditions of each nation. *Clean Energy Grid-Connected Technology Based on Smart Grid* by Li Peng et al. in the year 2011 (*Energy Procedia*) provides insights on the adoption of clean energy sources has been accelerated by the increased focus on managing energy scarcity, battling climate change, and reducing environmental pollution on a global scale. The imperative shift toward clean energy is driven by the necessity to reduce greenhouse gas emissions and transition toward sustainable energy alternatives. In response to these challenges, researchers have increasingly focused on integrating smart grid technology into clean energy research and development (R&D) initiatives. Smart grid technology plays a pivotal role in facilitating the efficient integration and utilization of clean energy resources within the existing energy infrastructure [18]. One key aspect of the synergy between smart grid technology and clean energy is the development of grid-connected control methods and techniques tailored to clean energy systems. These methods aim to optimize the integration of clean energy sources, such as solar and wind power, into the grid while ensuring stability, reliability, and efficiency [19].

Researchers in the year 2022 (*Energy Reports*) proposed the circular sustainable smart electric supply chain system. Four power-producing units in this system use both renewable and non-renewable energy sources to generate electricity. The power

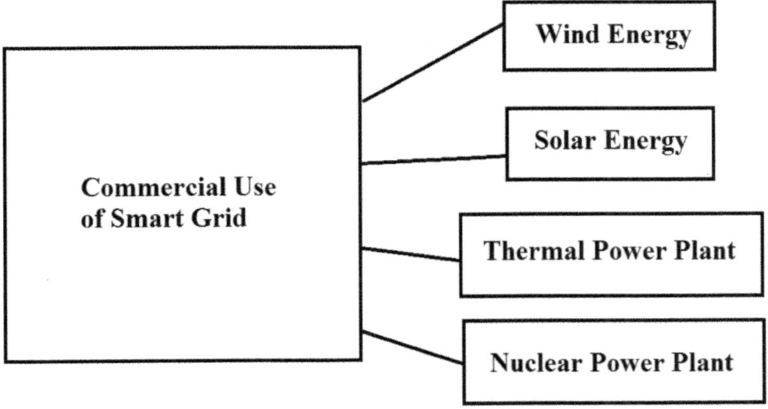

FIGURE 11.2 Smart grid management distributed energy resource assets [9].

allocation from various power generating units is managed by the smart grid management system, which also provides customers with electricity produced from renewable energy sources rather than non-renewable energy sources. To determine the ideal circumstances for the circular sustainable smart electric supply chain system to make the most profit in each of the two scenarios, an algorithm has been created. In the linear instance, the system makes more money than it does in the other. In order to comprehend the variation in profit, emission cost, and waste cost with fluctuations in certain parameters, a sensitivity analysis is conducted for the linear scenario. By investing optimally in green and waste minimization technology, the system attains higher profits. A larger profit will arise from the prudent distribution of power generation from each unit and the maximum amount of electricity produced by renewable power units [20].

The *Renewable and Sustainable Energy* Reviews assessment from 2022 attempted to look closely at the technology required to put a Behind the meter (BTM) system in place in this article. A critical component that was subsequently scrutinized was the impact of diverse metering and billing methods on the profitability of BTM systems. A thorough taxonomy of BTM energy storage system (ESS) services and applications has been supplied in order to investigate the opportunities and benefits that these systems provide. Though succinctly, every optimization strategy employed to address the energy management problems in the presence of BTM ESSs has been looked at and is explained in length. Additionally, the research provides insights into the role and significance of small-scale BTM ESSs in putting the concept of a smart grid powered by economical and sustainable resources [21].

The *International Journal of Electrical Power & Energy Systems* proposed a dynamic multi-stage model in 2023 for the coordinated expansion planning of transmission and distribution systems integrated with smart grid technologies. This model aims to coordinate the planning of transmission and active distribution system expansion with respect to short-term operational details. The results of Distribution System Operators (DSO)'s modeling of demand response program (DRP) in connection to Transmission system operators Locational marginal prices (TSO LMPs) in interface vehicles demonstrate DRP's ability to reduce overall operating and planning costs. After analyzing the operational hours of Electric Vehicle Control Systems (EVTCSs) in potential service zones, almost precise results were obtained for the Electric Vehicle Technology (EVT) charging pattern forecasting. It is not possible to completely understand how smart grid technologies affect distribution and transmission networks until operational considerations are included. The various difficulties of the suggested coordinated planning model with its large number of decision factors can be handled by the suggested Behavior Driven Development (BDD) algorithm [22].

Another research paper in the year 2016 (*Renewable and Sustainable Energy Reviews*) covers the latest issues in power line, wireless, cellular, virtual private networks, and information security. The three primary research issues that remain unresolved are quality of service support, transport protocols, and routing protocols. Comprehensive coverage is also given to new issues with intelligent electronics, demand side management, distribution, and energy management systems. The main problems with smart meters are those related to its design, implementation, and maintenance; how to develop demand management and control strategies; uncertainties about the policies of utility companies and the relevant governments regarding smart meters; flow-based algorithms for congestion control that reduce peak loads and thus reduce power congestion; cloud computing methods that offer incredibly precise forecasting; and broadcast and multi-agent control systems that enable distributed system automation [22].

The recent research in the year 2021, a recent publication, reported on advancements and research in the field of smart grids over the last 10 years [23]. The study indicates that the smart grid technology's idea and virtual reality stages have given way to the implementation phase. Scholarly interest in smart grids is evident from the 26,668 unique research articles that have been published in the last 10 years. The field of smart grid research has grown significantly during the past 5 years. This will soon alter the landscape for more adaptable and effective electrical power distribution. Given that the current grid's new age is only getting underway, there is still plenty of opportunity for enhancement and implementation of this concept [23]. We don't yet know how much research will be required to fully implement the smart grid concept, but recent advancements in smart grid technologies—like big data, demand side management, smart meters, and self-healing systems—are encouraging. This article, which reviewed smart grid technologies in 2016 (*Renewable and Sustainable Energy Reviews*), described how the need to modernize the electrical infrastructure prompted the creation of the smart grid. Eventually, the traditional grid became too small and needed more features. It has identified the characteristics and features of smart grids. In this article, the foundations of smart grid technology and related technologies have been covered, along with the research initiatives, challenges, and worries that have been surrounding them. Prospects for future research include cloud computing, battery systems, realistic large-scale integration of renewable energy sources, time series forecasting in smart grids, dependability and power quality investigations, and power flow optimization. Although the smart grid project involves spending money, time, and resources on continuing research and testing, it provides insights on reference-level experiences and technology. If substantial research is done in this field, the smart grid can help achieve energy sustainability and environmental conservation and preservation more successfully. Even while it could be challenging to predict the precise future of the smart grid, current developments show a dynamic blending of communities, mechanics, and sectors [24].

Dynamic Characteristics of Smart Grid Technology Acceptance discussed in detail in the year 2017 (*Energy Procedia*) highlights the significance of controlling knowledge of the hazards connected to smart grids, since perceived risk has a negative impact on intention to use and becomes a dependent variable as smart grid usage grows. Customers are aware of the advantages and the hazards, which include physical, functional/economic, and social/psychological concerns. It is necessary to make an effort to guarantee that consumer satisfaction matches expectations on the advantages of smart grids in order to reduce the discrepancy between expectations and satisfaction. Information that has been overemphasized shouldn't be shared during Public Relations (PR) and education. Maintaining a balance between increasing advantages and lowering hazards is necessary to encourage the spread of smart grids. There will be an increase in demand for associated technologies as smart grid usage increases; hence, technological development should concentrate on enhancing perceived ease of use by enhancing compatibility with existing technology and designing intuitive user-friendly interfaces [25].

Overall, the above article underscores the importance of international collaboration and knowledge sharing in advancing smart grid technologies worldwide. By leveraging the experiences and successes of pioneering nations, developing countries can accelerate their smart grid initiatives and contribute to the global transition toward more sustainable and efficient energy systems.

11.3 APPROACHES ON WASTE MANAGEMENT

The collection, handling, processing, and disposal of waste materials in a safe and environmentally responsible manner are referred to as "waste management." Numerous waste management techniques can improve sustainability and lessen the harm that rubbish does to the environment. The integration of garbage management and grid technologies marks a major shift in the direction of more efficient and sustainable waste management. Originally developed to optimize energy systems, smart grid technologies are already revolutionizing waste management practices. This integration uses real-time data, internet of things (IoT) sensors, and predictive analytics to transform garbage monitoring, collection, and processing into dynamic, data-driven operations. This paradigm shift has made it feasible to make proactive decisions. Waste management systems may be instantly adjusted to optimize routes, enhance recycling initiatives, and reduce environmental impact. Building smarter, greener, and more resilient communities may be facilitated by the synergy between waste management and grid technology, as we investigate this transformational fusion. Promoting sustainability and reducing the negative effects of trash on the environment need effective waste management. Waste management may be done in several ways, such as source separation, composting, landfilling, reusing, recycling, and decreasing. Combining these strategies can reduce waste production, protect natural resources, and advance a more sustainable future [1,7,10,11].

11.3.1 Local Prediction

1. Schedules for Dynamic Collections: Thanks to smart grid technology, waste management systems can now dynamically alter collection schedules in response to real-time data from sensors and smart bins. This improvement reduces costs, burns less gasoline, and ensures more efficient rubbish pickup [12–15].
2. Data-Based Perspectives: By giving waste management authorities access to useful information about regional waste trends, smart grids enable them to make well-informed judgments. This data is used to inform predictive analytics, which helps to forecast peak waste times and improves infrastructure design and resource allocation.
3. Integrating with Smart City Initiatives: Waste management grid technologies align with more extensive smart city initiatives. Local grids may be able to develop a holistic plan for urban sustainability and resilience by combining waste management data with other smart city applications like traffic control or environmental monitoring [9,17,18–20].
4. Emergency Situation Resilience: Grid technologies improve local waste management systems' resilience, especially in times of emergency or natural catastrophe. Authorities may prepare ahead and allocate resources more effectively during such situations by using predictive analytics to assist them anticipate higher trash creation [21–23].

11.4 GLOBAL PLANNING

The use of grid technology in waste management is the cutting edge of sustainable urban planning worldwide. As the world battles with mounting waste concerns, smart grids and cutting-edge technologies are being intentionally employed to improve trash management systems. Global planning makes use of smart sensors, IoT devices, and advanced data analytics to monitor, assess, and enhance the waste collection, disposal, and recycling processes. This comprehensive approach aims to reduce environmental impact, increase operational efficiency, and support the circular economy [20–26].

Cities and governments all around the world are creating comprehensive plans that leverage grid technology to transform trash management into a data-driven, dynamic industry. This global effort requires real-time garbage bin monitoring, efficient collection routes, and predictive analytics for trash-generating tendencies. Using renewable energy sources in waste treatment facilities guarantees that waste management operations have a lower carbon footprint, which is in line with sustainable standards. This global planning acknowledges the significance of waste management using grid technologies in promoting resilience and adaptation to the changing demands of rising urban populations, in addition to minimizing environmental effects [23–26]. The common goal of creating cities that are smarter, greener, and more efficient highlights how disruptive grid technologies have the potential to be in the development of global waste management systems. Reducing the carbon footprint of waste management activities is ensured by the incorporation of renewable energy sources into waste treatment facilities, which further aligns with sustainable practices. Global prediction models have a big impact on how well grid technologies work for waste management. Waste management authorities may enhance operational efficiency, optimize resource usage, and make well-informed judgments on waste management plans by employing distributed computing resources and producing precise forecasts of trash generation trends. We may anticipate more developments in global prediction models and grid technologies as waste management continues to grow, which will improve the efficiency and sustainability of waste management techniques globally [24–26].

Furthermore, cross-border partnerships and knowledge-sharing networks are encouraging the sharing of creative ideas and best practices. In addition to reducing environmental impact, this worldwide planning recognizes the importance of waste management through grid technologies in fostering resilience and adaptation to the changing demands of growing urban populations. The common goal of creating cities that are smarter, greener, and more efficient highlights how disruptive grid technologies have the potential to be in the development of global waste management systems. Not to add, energy resource management and waste reduction might undergo a radical change, thanks to smart grid technology. By utilizing advanced sensors and communication technologies, smart grid systems may get up-to-date information on energy use and demand. Operators can reduce waste and optimize energy distribution as a result. Smart grids can also make it easier to integrate renewable energy sources, including wind and solar electricity, into the system. This might reduce carbon emissions and the need for fossil fuels. Ultimately, this can reduce the amount of waste produced and mitigate the effects of climate change. When everything is said and done, smart grid technologies hold enormous potential for improving waste management in the energy sector [5–7,10].

11.4.1 Wide Area Monitoring for Distribution and Transmission

1. Enhanced Contextual Awareness: Wide area monitoring systems give grid operators access to real-time data on the operating status of distribution and transmission networks. Operators use sophisticated sensors and synchronized phasor measuring units (PMUs) to gather exact voltage, current, and phase angle data across large areas. This improved situational awareness makes it easier to detect potential issues early on, which lowers the risk of cascading failures and significantly improves grid reliability overall [1,2,8,27].
2. Control and Stability of the Grid: Dynamic stability evaluations of the power system are made easier by the synchronized and fast data from wide area monitoring. Operators can identify oscillations and disturbances via real-time phasor measurement analysis. This allows for prompt control steps to preserve grid stability. This involves modifying the output of the generator, turning on compensators, or rearranging the network, offering a proactive approach to prevent system-wide disruptions [2–6].
3. Predictive Maintenance and Early Failure Detection: Wide area monitoring systems are excellent at identifying faults early on since they continuously

check the state of grid equipment and transmission lines. Utilities can employ predictive maintenance solutions by identifying deviations from typical operating conditions. Proactive maintenance reduces operating expenditures over time by improving system reliability, extending the lifespan of critical infrastructure components, and reducing downtime [6,7,10,11].

4. Combining Sustainable Energy Sources: As the renewable energy sources are becoming more and more integrated, broad area monitoring is essential to managing their intermittent and dispersed character. Because of technology, grid managers can monitor and operate it more flexibly, taking into account variations in the production of renewable energy. This supports the worldwide shift to a more sustainable and renewable energy source by ensuring a steady and dependable power supply [12–15].
5. Cybersecurity: Because wide area monitoring systems are vulnerable to cyberattacks, they primarily rely on digital communication and control systems. Cybersecurity measures are used to prevent unauthorized access, data breaches, and potential power outages. Strict access controls, safe communication channels, and encryption techniques must be put in place as a result. Data is encrypted during transmission between devices and control centers to ensure data integrity and confidentiality. This prevents malicious actors from intercepting or changing crucial information [9,16–19].

11.4.2 Factors Affecting Waste Management and Energy Policy

Plans for waste management and energy policies are impacted by a number of factors, which affect the efficient and sustainable use of resources. Public opinion, technological advancements, environmental repercussions, economic concerns, and regulatory frameworks are all important factors. Innovations in grid technology, the concept of a circular economy, and the integration of renewable energy sources are becoming more significant factors in the decision-making process. Finding a balance between social welfare, economic growth, and environmental preservation is a difficult but necessary step in creating effective waste management and energy policy. Moreover, grid technologies provide increased fault tolerance and dependability. Grid systems that have dispersed resources can reroute workloads to other resources that are accessible to resolve failures or interruptions in specific components. Because of their high availability and low downtime, grids are appropriate for mission-critical applications that must run continuously. At the national level, energy efficiency evaluation has several uses. It is helpful in identifying areas for improvement and the primary strategies for reducing energy losses, conserving and wisely using finite energy sources, lowering carbon and greenhouse gas emissions, and so forth [20–23].

In light of the worldwide endeavors to tackle climate change and transition to a more sustainable future, the assessment of energy efficiency has gained significant prominence. The International Energy Agency (IEA) estimates that energy efficiency initiatives can provide around 40% of the energy-related reductions in greenhouse gas emissions required to achieve global climate targets. Furthermore, energy efficiency may lower consumer energy bills, boost corporate competitiveness, and generate employment in the clean energy industry [24–26].

A nation's energy policies have a major role in determining its overall energy landscape. Governments and politicians set the course of energy development, considering factors including energy security, environmental sustainability, and economic feasibility. The worldwide move toward renewable energy sources demonstrates a commitment to reducing carbon emissions and mitigating the consequences of climate change. Energy-related forward-looking policies increasingly include grid modernization, reducing energy consumption, and the integration of smart technologies. International cooperation and agreements also play a major role in building the adoption of sustainable energy technologies worldwide [19].

1. Economic Factors: Economic factors have a big impact on waste management and energy policy. The cost of energy generation and waste disposal can affect policy decisions about recycling, garbage reduction, and the use of renewable energy sources. Economic incentives such as taxes, subsidies, and feed-in tariffs can influence policy decisions by either promoting or inhibiting the adoption of specific technology or practices [21].
2. Technological Factors: Innovations in technology have the potential to impact policy modifications related to waste and energy management. Emerging technologies like carbon capture and storage, waste-to-energy conversion, and smart grid systems may have an impact on policy decisions concerning waste management, renewable energy, and grid stability. The adoption and implementation of policies may also be impacted by the affordability and accessibility of these technologies [22].
3. Environmental Factors: Energy and waste management policies are heavily influenced by environmental concerns. Environmental concerns that might affect policy decisions about waste management and energy production include resource depletion, air and water pollution, and climate change. Environmental concerns frequently serve as the driving force behind policies intended to lower greenhouse gas emissions, enhance the quality of the air and water, and encourage the sustainable use of resources [26].
4. Social Variables: Public opinion, cultural norms, and demographic changes are examples of social variables that might influence energy and waste management policies. Policy choices about environmental preservation, renewable energy, and recycling can be influenced by public support for these causes. The adoption of innovative technology and energy- and waste-saving practices can also be influenced by social considerations [24].
5. Political Aspects: International agreements, governmental objectives, and legislative frameworks are just a few examples of the political factors that can have a big impact on a nation's waste management and energy policy. Government initiatives aiming at waste reduction, energy efficiency promotion, and the development of renewable energy sources are regularly impacted by political interests and objectives. International agreements such as the Paris Climate Change Agreement might impact policy decisions related to energy and waste management.
6. Infrastructure Factors: Infrastructure issues also have an impact on waste management and energy policy. Policy choices on the production of energy

and waste management can be influenced by the capacity and availability of recycling facilities, waste management facilities, and energy infrastructure. Policy choices regarding waste and energy can also be influenced by the construction of new infrastructure, such as waste-to-energy facilities, renewable energy installations, and electric car charging stations [25].

11.5 KEY OBSERVATIONS OF ENERGY POLICY

Energy policies usually place a strong emphasis on diversifying energy sources to increase energy security and reduce dependency on a single source. This might include promoting the use of renewable energy sources like solar, wind, hydro, and geothermal energy in addition to traditional fossil fuels. Policies place a strong emphasis on energy efficiency and conservation in order to maximize energy use across sectors. This means creating policies for appliances, endorsing energy-efficient equipment, and devising strategies to reduce energy waste. With the rising focus on climate change on a global basis, energy policies increasingly include initiatives to reduce greenhouse gas emissions. This might mean setting targets for cutting emissions, encouraging the use of cleaner technologies, and quickening the transition to renewable and low-carbon energy sources. To promote innovation in the energy industry, governments frequently fund R&D projects. In order to take advantage of new possibilities and challenges, this involves promoting technological breakthroughs, energy storage, and innovative energy solutions. Incentives and regulatory frameworks to direct market behavior are examples of energy strategies. This may entail putting in place carbon trading schemes, setting feed-in tariffs for renewable energy, and offering financial rewards for energy-saving behaviors [23–26].

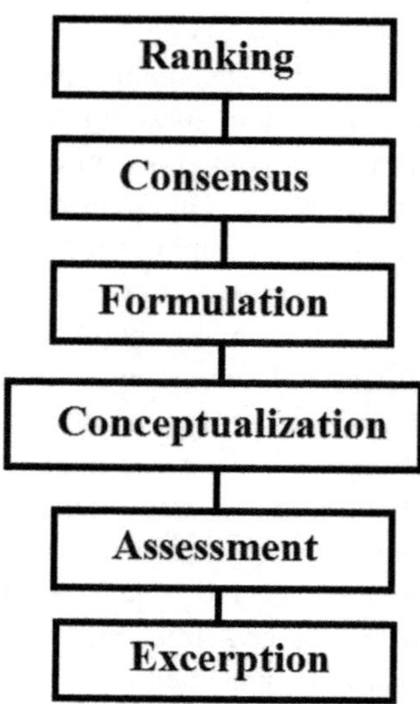

FIGURE 11.3 Steps for energy policy development [26].

Spreading Your Energy Sources: To increase energy security and reduce dependency on a single supplier, energy policy generally emphasizes the diversification of energy sources. This might include promoting the use of renewable energy sources like solar, wind, hydro, and geothermal energy in addition to traditional fossil fuels.

Energy Efficiency and Conservation: Policies place a strong emphasis on energy efficiency and conservation measures in order to maximize energy use across sectors. This means creating policies for appliances, endorsing energy-efficient equipment, and devising strategies to reduce energy waste.

Mitigation of Climate Change and Sustainability of the Environment: Energy strategies are progressive, including steps to lower greenhouse gas emissions in light of the increased attention being paid to climate change on a worldwide scale. This might entail establishing goals for reducing emissions, supporting the use of greener technology, and accelerating the switch to renewable and low-carbon energy sources.

R&D Initiatives: To promote innovation in the energy industry, governments frequently sponsor R&D projects. Energy storage and innovative energy solutions need to be encouraged in order to take advantage of the opportunities and challenges brought about by recent advancements.

Control and Promotion of the Market: Using incentives and using legal frameworks to shape consumer behavior are two examples of energy tactics. This might entail putting in place feed-in tariffs for renewable energy, creating carbon trading schemes, and offering incentives to people who use less energy.

International Collaboration: An essential part of the energy strategy is international collaboration. Its main objective is to encourage international cooperation and coordination on matters about energy, such as technical advancement, energy security, and climate change. Participation in international accords, like the Paris Agreement, and diplomatic initiatives to advance energy security and sustainability are examples of international cooperation strategies [14].

Energy Infrastructure: Energy infrastructure is a crucial component of energy strategy. Ensuring that energy is delivered to customers in a safe, dependable, and efficient way is its primary objective. Energy infrastructure policies include, among other things, investments in transmission and distribution networks, smart grid technologies, and distributed energy resources [17].

11.5.1 Benefits on Waste Management

Advantages of Waste Management: Effective garbage management has a number of benefits that go beyond just getting rid of waste. Environmental pollution is decreased when hazardous pollutants are released into the air, land, and water in less quantities, thanks to a well-operating waste management system. Implementing recycling and resource recovery strategies also helps to reduce the demand on finite resources, develop a circular economy, and preserve valuable raw materials. Waste-to-energy systems promote sustainable power generation and reduce dependency on conventional energy sources. Appropriate waste management also enhances public health as it lowers the possibility of disease transmission. Involving communities in waste reduction and recycling initiatives also helps to promote social awareness and build a shared responsibility for environmental care. Taking everything into account, prudent waste management is essential to protecting ecosystems, saving resources, and promoting development [19–22]:

1. Protection of the Environment: Environmental protection is one of the main advantages of waste management. We can stop soil, water, and air pollution, as well as lower greenhouse gas emissions that fuel climate change by disposing of garbage appropriately.
2. Preservation of Natural Resources: Waste management also helps to preserve natural resources by reducing the need for raw materials. Reusing and recycling materials like paper, plastics, and metals may reduce the use of new resources and save energy.
3. Public Health: To ensure public health, effective waste management is required. Not only may incorrect garbage disposal attract rodents and vermin, but it can also transmit pathogens and illnesses.
4. Energy Production: Waste management is an additional facet of energy production. Utilizing waste-to-energy technologies, such as gasification and incineration, which may convert waste materials into heat and power, can reduce the amount of fossil fuels used.
5. Financial Gains: There might be financial benefits from efficient garbage management. Reusing and recycling materials may strengthen local economies, create jobs, and help governments and businesses save money on disposal expenses.

11.6 CONCLUSION

The integration of smart grid technology into waste management represents a substantial step forward in the direction of a more advanced, sustainable, and effective approach. Smart grids provide real-time data, predictive analytics, and connection to waste management systems, allowing them to maximize collection, reduce running costs, and enhance overall efficiency. Because of their dynamic adaptability, smart grids provide for better resource allocation and less environmental impact by enabling proactive responses to changing waste patterns. By propelling data-driven decision-making, smart grid technologies help move waste management away from outdated procedures and toward a more resilient and innovative model. In addition to streamlining operations, the combination of smart grids and garbage disposal promotes a broader paradigm shift toward more intelligent, sustainable, and eco-friendly urban settings.

In conclusion, grid technologies are critical to the state of the energy industry today since they provide a number of benefits and opportunities for a more stable and sustainable energy system. The way we generate, transmit, and utilize energy has been drastically altered as a result of the advancements in grid technology. Additionally, it has increased grid stability, made it possible for renewable energy sources to be integrated more deeply, and given customers more influence over the energy market. Grid technologies empower consumers by giving them more control and versatility over how much energy they use. Customers can utilize smart meters, energy management systems, and DRPs to actively regulate their energy usage by optimizing it based on price signals and grid conditions. This not just enhances the overall stability and efficiency of the grid but also lowers customer energy bills.

Grid technologies also make it possible to create new business models and prospects in the energy industry, in addition to these advantages. With the use of cutting-edge grid technology, distributed energy resources like electric car charging stations and rooftop solar panels may be easily included in the system. The potential of grid technology to make the integration of renewable energy sources easier is one of its many noteworthy benefits. Grid technologies like ESSs and smart grids make it possible to manage and use intermittent renewable energy more effectively as solar and wind power become more widely used. It is important to remember that supportive investments and policies are needed for grid technologies to continue to improve and be deployed. Governments and regulatory agencies must provide advantageous policies, such as feed-in tariffs, net metering, and financing for R&D, to encourage the use of grid technology. Furthermore, the effective use and advancement of grid technologies depend on collaboration between stakeholders, including providers of technology, consumers, and utilities. By enabling the integration of renewable energy, enhancing grid stability, giving consumers more power, and building a more resilient and sustainable energy system, grid technologies have the potential to completely transform the energy sector. Grid technology, with the right investments, policies, and collaboration, might pave the way for a cleaner, more efficient, and decentralized energy future. In order to address concerns related to energy security, climate change mitigation, and sustainable development, grid technology must be completely embraced and utilized.

REFERENCES

1. Nižetić, S., Djilali, N., Papadopoulos, A., & Rodrigues, J. J. (2019). Smart technologies for promotion of energy efficiency, utilization of sustainable resources and waste management. *Journal of Cleaner Production*, 231, 565–591.
2. Kinally, C., Antonanzas-Torres, F., Podd, F., & Gallego-Schmid, A. (2022). Off-grid solar waste in sub-Saharan Africa: Market dynamics, barriers to sustainability, and circular economy solutions. *Energy for Sustainable Development*, 70, 415–429.
3. Lu, W., Lee, W. M., Bao, Z., Chi, B., & Webster, C. (2020). Cross-jurisdictional construction waste material trading: Learning from the smart grid. *Journal of Cleaner Production*, 277, 123352.
4. Vincent, E. N., & Yusuf, S. D. (2014). Integrating renewable energy and smart grid technology into the Nigerian electricity grid system. *Smart Grid and Renewable Energy*, 5, 220–238.
5. Mugendi, K. D., Mireri, C. O., & Enevoldsen, M. K. (2023). Towards development of effective policies and regulations for sustainable off-grid solar electronic waste management systems in Kenya. *Engineering Reports*, 6, e12805.
6. Bhattarai, T. N., Ghimire, S., Mainali, B., Gorjian, S., Treichel, H., & Paudel, S. R. (2023). Applications of smart grid technology in Nepal: Status, challenges, and opportunities. *Environmental Science and Pollution Research*, 30(10), 25452–25476.
7. Neffati, O. S., Sengan, S., Thangavelu, K. D., Kumar, S. D., Setiawan, R., Elangovan, M., & Velayutham, P. (2021). Migrating from traditional grid to smart grid in smart cities is promoted in developing countries. *Sustainable Energy Technologies and Assessments*, 45, 101125.

8. Bayindir, R., Colak, I., Fulli, G., & Demirtas, K. (2016). Smart grid technologies and applications. *Renewable and Sustainable Energy Reviews*, 66, 499–516.
9. Tuballa, M. L., & Abundo, M. L. (2016). A review of the development of smart grid technologies. *Renewable and Sustainable Energy Reviews*, 59, 710–725.
10. Seo, S., Aramaki, T., Hwang, Y., & Hanaki, K. (2003). Evaluation of solid waste management system using fuzzy composition. *Journal of Environmental Engineering*, 129(6), 520–531.
11. Winkler, J., & Bilitewski, B. (2007). Comparative evaluation of life cycle assessment models for solid waste management. *Waste Management*, 27(8), 1021–1031.
12. Adam, M., Okasha, M. E., Tawfeeq, O. M., Margan, M. A., & Nasreldeen, B. (2018). Waste management system using IoT. In *2018 International Conference on Computer, Control, Electrical, and Electronics Engineering (ICCCEEE)*, Khartoum, Sudan (pp. 1–4).
13. Dubey, S., Singh, P., Yadav, P., & Singh, K. K. (2020). Household waste management system using IoT and machine learning. *Procedia Computer Science*, 167, 1950–1959.
14. Salehi-Amiri, A., Akbapour, N., Hajiaghaei-Keshteli, M., Gajpal, Y., & Jabbarzadeh, A. (2022). Designing an effective two-stage, sustainable, and IoT based waste management system. *Renewable and Sustainable Energy Reviews*, 157, 112031.
15. Pardini, K., Rodrigues, J. J., Kozlov, S. A., Kumar, N., & Furtado, V. (2019). IoT-based solid waste management solutions: A survey. *Journal of Sensor and Actuator Networks*, 8(1), 5.
16. Chaudhari, M. S., Patil, B., & Raut, V. (2019). IoT based waste collection management system for smart cities: An overview. In 2019 3rd *International Conference on Computing Methodologies and Communication (ICCMC)*, Erode, India (pp. 802–805).
17. Anuradha, D., Vanitha, A., Priya, S. P., & Maheshwari, S. (2017). Waste management system using IoT. *International Journal of Computer Science Trends and Technology (IJCST*), 5(2), 152–155.
18. Ardito, L., Procaccianti, G., Menga, G., & Morisio, M. (2013). Smart grid technologies in Europe: An overview. *Energies*, 6(1), 251–281.
19. Tao, F., Zhang, L., & Nee, A. Y. C. (2011). A review of the application of grid technology in manufacturing. *International Journal of Production Research*, 49(13), 4119–4155.
20. Colak, I., Sagiroglu, S., Fulli, G., Yesilbudak, M., & Covrig, C. F. (2016). A survey on the critical issues in smart grid technologies. *Renewable and Sustainable Energy Reviews*, 54, 396–405.
21. Kim, S. C., Ray, P., & Reddy, S. S. (2019). Features of smart grid technologies: An overview. *ECTI Transactions on Electrical Engineering, Electronics, and Communications*, 17(2), 169–180.
22. Harighi, T., Bayindir, R., Padmanaban, S., Mihet-Popa, L., & Hossain, E. (2018). An overview of energy scenarios, storage systems and the infrastructure for vehicle-to-grid technology. *Energies*, 11(8), 2174.
23. Ekanayake, J. B., Jenkins, N., Liyanage, K. M., Wu, J., & Yokoyama, A. (2012). Smart Grid: Technology and Applications. John Wiley & Sons, New York.
24. Brunner, P. H., & Fellner, J. (2007). Setting priorities for waste management strategies in developing countries. *Waste Management & Research*, 25(3), 234–240.
25. Kansal, A. (2002). Solid waste management strategies for India. *Indian Journal of Environmental Protection*, 22(4), 444–448.
26. Ali, T., Al Mansur, A., Shams, Z. B., Ferdous, S. M., & Hoque, M. A. (2011, September). An overview of smart grid technology in Bangladesh: Development and opportunities. In *2011 International Conference & Utility Exhibition on Power and Energy Systems: Issues and Prospects for Asia (ICUE)*, Pattaya, Thailand (pp. 1–5).
27. Samad, T., & Kiliccote, S. (2012). Smart grid technologies and applications for the industrial sector. *Computers & Chemical Engineering*, 47, 76–84.

12 Advanced Emission Control Techniques Used in Foundries for Separating Contaminants from the Working Medium

A. Vimal and S. Venkatesh

12.1 INTRODUCTION

In the view of the public, the potential source of air pollution is foundry industries. The pollution from industries affects the surrounding areas in terms of property, vegetation, animals, and even humans. Humans are affected by respiratory diseases, while animals are affected by their well-being, the growth of crops is reduced, and the corrosion of minerals due to the emission of sulfur dioxide and other acid gases. The following are many sources of pollutants.

1. Effluents from dust and furnaces.
2. Solid particles and suspended matters from fluid streams.
3. Dust and ash from exhaust gases.
4. CO_2 emission from power plants.
5. Flue-gas condensation.
6. Smoke and fly ash from the combustion chamber.
7. Odor and gas compounds.

In order to separate these contaminants from the working medium, the foundryman must under the cause of the pollution, the source of the pollutants, and techniques to measure and control the emissions. This is the key parameter discussed in this chapter along with the working, roles, design, and investigation of the advanced emission control techniques.

DOI: 10.1201/9781003452072-12

12.2 INERTIAL SEPARATOR (CYCLONE)

An inertial separator, commonly known as a cyclone separator, is a device used for the separation of solid particles from a gas or liquid stream based on their inertial forces. It utilizes centrifugal force to separate particles with different sizes or densities from the carrying fluid [1] (Figure 12.1).

12.2.1 Working of an Inertial Separator

1. Mixture Introducing: The cyclone separator is fed tangentially with a gas or liquid mixture containing particles.
2. Centrifugal Force Generation: Due to the tangential feed, a spiral motion is formed that generates centrifugal force.
3. Particle Separation: The centrifugal force causes greater inertia to the heavier particles due to which it moves toward the cyclone wall, whereas the lighter particles move in the direction of the gas or liquid flow.
4. Particle Collection: The heavier particles striking the wall of the cyclone lose kinetic energy and, due to gravitational force, settle down at the bottom of the cyclone, while the gas or liquid, free from heavier particles, gets collected from the top of the cyclone.

12.2.2 Roles of Cyclone Separator in Industries

1. Dust Collector: Cyclone acts as a dust collector by removing the dust and particulate matter from air streams. Industries such as mining, carpentry, and sheet metal widely use cyclones to maintain air quality and satisfy environmental regulations.

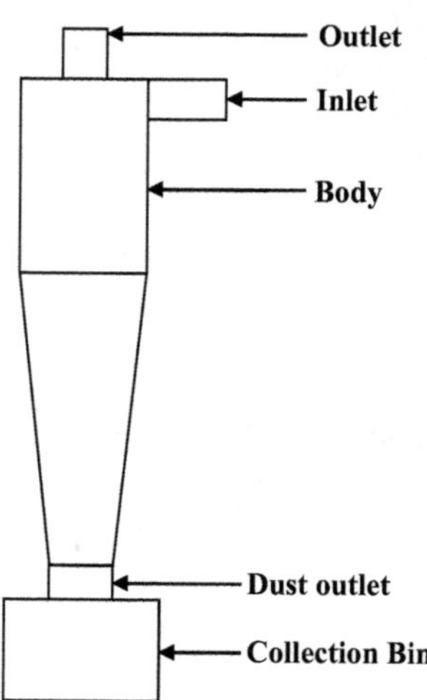

FIGURE 12.1 Schematic diagram of the cyclone.

2. Gas Collector: Cyclone behaves like a gas collector by cleaning the gases from solid particles. It complies with the emission standard and also meets the clean gas requirement in industrial applications.
3. Pre-Filter: Cyclone serves as a pre-filter when combined with other filtering processes. It removes high-density particles, thereby enhancing the efficiency of the downstream filters.
4. Particle Classifier: Cyclone functions as a classifier, separating particles based on their size and density. In cement industries, sands of different types are segregated by the use of particle classifiers.

12.2.3 Experimental and Numerical Investigation

The research paper by Wang et al. [1] presents an experimental and numerical investigation on the separation of hydrophilic fine particles using a novel technique called heterogeneous condensation preconditioning in gas cyclones. The study aims to enhance the separation efficiency of fine particles with diameters ranging from 0.2 to 2 μm in a gas cyclone by using a condensation technique.

The paper discusses the design and construction of a novel experimental setup to investigate the separation efficiency of hydrophilic particles. The setup uses a single-stage gas cyclone with a 300-mm diameter and a 1600-mm height, and a condensation chamber to precondition the incoming gas flow. The study also includes a numerical simulation of the gas flow and particle behavior inside the gas cyclone using the computational fluid dynamics (CFD) method.

The results of the experiments and numerical simulations demonstrate that the heterogeneous condensation preconditioning technique can significantly improve the separation efficiency of hydrophilic fine particles in a gas cyclone. The technique is found to be effective in reducing the particle size distribution and increasing the separation efficiency by up to 30%. The numerical simulation also provides insights into the flow patterns and particle trajectories inside the gas cyclone.

12.2.4 Performance Enhancement of Cyclone Separator

Jafarnezhad et al. [2] investigate enhancing the performance of a cyclone separator under high-temperature operating conditions by utilizing different shapes of the vortex finder. The study addresses the challenges of efficient particle separation in environments with elevated temperatures, which are often encountered in industrial applications.

The paper presents a comprehensive investigation into the effect of vortex finder shapes on the performance of the cyclone separator. As the inlet temperature increases, the swirling flow across the cyclone gets weaker leading to decrease in separation efficiency. Through experimental and numerical analysis, Jafarnezhad evaluates various vortex finder geometries and their impact on key performance parameters, including separation efficiency and pressure drop.

The research highlights the significance of the vortex finder design in improving the cyclone separator's performance under high-temperature conditions. It discusses the advantages and limitations of different vortex finder shapes like base, divergent vortex finder, and convergent vortex finder. The simulation of the flow field of different shapes was made using RSM (Reynold Stress Model) to enhance particle separation while minimizing energy losses.

The experimental and numerical results demonstrate that convergent vortex finder shapes lead to improved separation efficiency and reduced pressure drop in high-temperature operating conditions. Jafarnezhad provides valuable insights into the underlying mechanisms and flow patterns within the cyclone separator, aiding in the understanding of the observed performance enhancements.

12.3 ELECTROSTATIC PRECIPITATOR (ESP)

An electrostatic precipitator (ESP) is an emission control device widely used in various industries to remove fine particulate matter, such as dust and ash, from exhaust gases. It operates on the principle of electrostatic attraction to capture and remove particles from the gas stream [3] (Figure 12.2).

12.3.1 Working of an Electrostatic Precipitator

1. Particle Charging: The electrostatic precipitator has a series of plates, which act as a high-voltage electrode to create a discharge. The gas containing the particles is allowed to pass through these plates, thereby charging the particles.
2. Gas Ionization: The gas molecules in the vicinity are ionized by the discharge. As a result of ionization, free electrons and positive ions are created.
3. Electrostatic Attraction: The negatively charged particles are then attracted toward the positively charged electrodes, named as collection plates, which are grounded. This electrostatic attraction causes the charged particles to migrate toward and stick to the electrodes.
4. Particle Removal: The collected particles form a layer on the collection plates, which are removed from the system using any of the mechanical rapping.

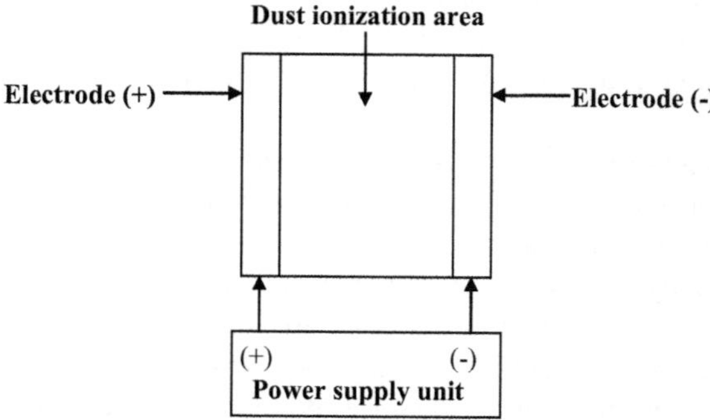

FIGURE 12.2 Schematic diagram of electrostatic precipitator and its components.

12.3.2 Roles of Cyclone Separator in Industries

1. Power Plants: ESPs are widely used in coal-fired power plants to capture the fly ash from the combustion of coal before the flue gases are released into the atmosphere.
2. Cement Industry: Cement manufacturing processes involve the production of exhaust gases, which contain dust and pollutants. ESPs support emission control by removing these pollutants from the exhaust gases.
3. Steel Mills: Steel production generates particulate matter in various stages of its process. These particulate matters are subjected to process through ESPs, thereby controlling emissions.
4. Pulp and Paper Mills: ESPs are applied to control emissions from processes like paper pulping, which releases particulate matter and other pollutants.
5. Chemical Processing Plants: Industries involved in chemical manufacturing assign ESPs to control emissions, thereby maintaining air quality.
6. Textile Industry: In textile industries, certain processes involve the production of airborne particles, where ESPs are employed for emission control.

12.3.3 Flow Simulation in Electrostatic Precipitator

Bhasker [4] provides an overview of ESPs and their role in controlling particulate matter emissions in industrial applications. It then introduces the concept of turning vanes, which are commonly used in ESP ducts to enhance the flow distribution and minimize turbulence.

Bhasker [4] describes the methodology employed in the research, focusing on the CFD modeling techniques used to simulate the flow patterns within the ESP ducts. The study aims to evaluate the effectiveness of turning vanes in improving flow uniformity and reducing pressure drop.

Throughout the paper, the author presents and analyzes the simulation results obtained from the CFD models. The findings provide valuable insights into the impact of turning vanes on flow characteristics, including velocity distribution, turbulence levels, and pressure drop.

The research paper discusses the implications of the simulation study, highlighting the benefits of utilizing turning vanes in ESP ducts. Bhasker [4] emphasizes the role of turning vanes in minimizing flow deviations, improving particle collection efficiency, and optimizing the overall performance of ESP systems.

By delivering a clear and concise presentation of the research methodology and results, "Flow Simulation in Electro-Static-Precipitator (ESP) Ducts with Turning Vanes" contributes to the understanding and optimization of ESP duct design. The study offers valuable guidance for engineers and researchers involved in the design and operation of ESP systems.

12.3.4 Development and Testing of Tube-Type Wet ESP

"Development and Testing of Tube-Type Wet ESP for the Removal of Particulate Matter and Tar from Producer Gas" by Parihar et al. [5] is a research paper that focuses on the development and testing of a tube-type wet ESP for the efficient removal of particulate matter and tar from producer gas.

The paper begins by highlighting the importance of controlling particulate matter and tar emissions from producer gas, which is generated in various industrial processes. Parihar et al. [5] introduce the concept of a tube-type wet ESP as an effective solution for the removal of these pollutants.

The author provides a detailed description of the design and construction of the wet ESP, including the arrangement of collection tubes, spray nozzles, and demisters. The paper emphasizes the use of water as a medium to enhance the capture and removal of fine particles and tar from the gas stream.

Parihar et al. [5] present the results of experimental testing conducted to evaluate the performance of the wet ESP. The findings demonstrate the efficiency of the system in removing particulate matter and tar, as well as its potential for reducing the emissions of these pollutants to acceptable levels.

The research paper discusses the implications of the study, highlighting the advantages of the tube-type wet ESP, such as its simplicity, reliability, and effectiveness in cleaning producer gas. Parihar et al. [5] also address the challenges and limitations associated with the wet ESP design, providing insights for further improvements.

12.4 CO_2 CAPTURING SYSTEM

A CO_2 capturing system is a technology that captures carbon dioxide (CO_2) emissions from various sources such as power plants or industrial processes, preventing them from being released into the atmosphere and contributing to climate change [6] (Figure12.3).

12.4.1 Working of CO_2 Capturing System

1. *Capture Phase:* a) Pre-Combustion Capture: In this phase, CO_2 is captured from flue gas before combustion. Hydrogen produced from fuel reacts with CO_2 to form water, which is then separated.
2. b) Post-Combustion Capture: This phase captures CO_2 from flue gases after combustion. The flue gases pass through a solvent, which absorbs CO_2,

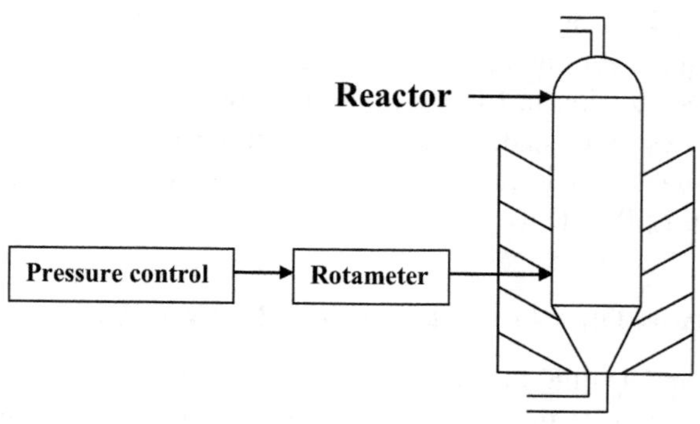

FIGURE 12.3 Test flow reactor schematic diagram.

forming a solution. This solution is then processed to separate and recover the captured CO_2.

3. Transportation Phase: The captured CO_2 needs to be transported to a suitable storage site. Depending upon the specific conditions and distance, the transportation phase involves pipelines, ships, or trucks.
4. Storage Phase: CO_2 is injected into geological formations such as non-mineable coal seams, deep saline formations, or depleted oil gas. The storage sites must have suitable characteristics to securely contain the CO_2 without leakage.

12.4.2 Roles of CO_2 Capturing System in Industries

1. Power Generation: CO_2 capturing systems can be integrated into fossil fuel power plants to reduce emissions. This is particularly relevant in regions where conversion to renewable energy sources is challenging.
2. Industrial Processes: In cement production, steel manufacturing, and chemical production, significant amounts of CO_2 are released. CO_2 capturing systems are involved to reduce the carbon footprint of these industries.
3. Natural Gas Processing: Natural gas consists of CO_2 and its removal is necessary to meet the pipeline standards. CO_2 capturing systems can be used to separate and capture CO_2 from natural gas streams.
4. Bioenergy and Biomass: In bioenergy facilities and biomass power plants, CO_2 capturing can be applied to address emissions associated with the combustion of organic materials.
5. Hydrogen Production: CO_2 capture is crucial in the production of hydrogen, especially when it involves natural gas reforming. The captured CO_2 prevents its release into the atmosphere.
6. Direct Air Capture (DAC): Direct air capture technology focuses on capturing CO_2 directly from the ambient air. This approach is explored as a means to address emissions from dispersed sources.

12.4.3 Carbon Dioxide Capture by Ammonium Hydroxide Solution

Hamouda et al.'s [6] research paper presents a compelling investigation into carbon dioxide (CO_2) capture using ammonium hydroxide solution and its potential application in the cement industry. The paper explores the efficiency and feasibility of this method, considering its effectiveness in reducing CO_2 emissions during cement production.

The author provides a clear and concise overview of the experimental setup and methodology, showcasing the rigorous scientific approach employed. The optimum parameter for maximum CO_2 removal rate is an ammonia concentration below 14%, a reaction temperature below 20°C, and a reaction time of 30 minutes. The results demonstrate promising CO_2 capture capabilities of ammonium hydroxide solution, offering insights into its potential as a viable solution for the cement industry.

The paper also discusses the practical implications and challenges associated with implementing this technology, including the cost implications, process integration, and environmental considerations. This technology can be implemented into the plant by considering plant design, utilities, and materials of construction. Hamouda et al. [6] provide thoughtful analysis and suggestions for future research and development, further enriching the value of the paper.

12.4.4 Advances in Carbon Capture and Storage

Shukla et al.'s [7] research paper on the advances of carbon capture and storage (CCS) in coal-based power generating units in the Indian context provides a valuable and comprehensive analysis of this critical technology. The paper examines the latest developments, challenges, and opportunities in implementing CCS to reduce CO_2 emissions from coal power plants in India.

Shukla et al. [7] adeptly outline the technical aspects of CCS, including various capture technologies and storage options, while considering their applicability within the Indian energy landscape. Among the capturing techniques, post-combustion capture is the most appropriate method to capture CO_2 gas. It incorporates solvent absorption, cryogenic separation, membrane partitioning, and stable sorbent adsorption. The captured CO_2 can be transported using high-weight pipelines for 1,000 km. The storage of the gas can be made both in land and seaward by infusing CO_2 in underground muddy bowls. The paper emphasizes the importance of CCS in addressing climate change and meeting India's growing energy demands sustainably.

This work contextually focuses on India, recognizing the unique challenges and opportunities specific to the country's coal-based power sector. Shukla et al. [7] offer insights into policy frameworks, economic considerations, and public acceptance, shedding light on the holistic implementation of CCS in India.

12.5 SCRUBBER

Scrubber system is an air pollution control equipment used to filter the industrial exhaust stream to be free from particulates. Scrubbers can also be utilized for flue-gas condensation, which recovers heat from hot gases [8] (Figure 12.4).

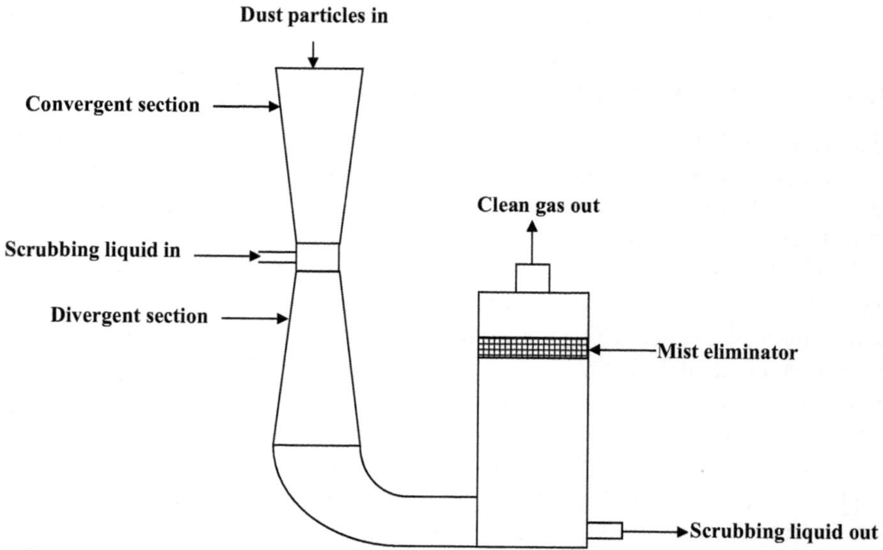

FIGURE 12.4 Venturi scrubber schematic diagram.

12.5.1 Working of Venturi Scrubber

1. The basic principle of a scrubber involves bringing the polluted gas into contact with a liquid, often water. This liquid captures or reacts with the pollutants, removing them from the gas stream.
2. In wet scrubbers, chemical reactions often occur between the pollutants and the liquid.
3. In dry scrubbers, solid material is used to capture pollutants. The solid material may be injected into the gas stream, and the pollutants adhere to the surface of the material.

12.5.2 Roles of Venturi Scrubber in Industries

1. Waste Incineration: Incineration of waste materials can release harmful gases. Scrubbers are employed to capture pollutants like hydrogen chloride, sulfur compounds, and particulate matter from the incinerator exhaust.
2. Oil Refineries: Refineries use scrubbers to control emissions of sulfur compounds, hydrogen sulfide (H_2S), and other pollutants generated during various refining processes.
3. Semiconductor Manufacturing: In the semiconductor industry, scrubbers are used to remove hazardous gases and chemicals used in the manufacturing processes to maintain a clean and safe working environment.
4. Mining Operations: Scrubbers can be used in mining operations to control emissions of dust and other pollutants associated with mining activities.

12.5.3 Self-Priming Venturi Scrubber

The research paper titled "Study of Hydrodynamics and Iodine Removal by Self-Priming Venturi Scrubber" presents a comprehensive investigation into the hydrodynamic characteristics and iodine removal efficiency of a self-priming venturi scrubber. Ahad et al. [9] contributed to the understanding and optimization of venturi scrubber technology.

Ahad et al. [9] provide a clear introduction, highlighting the importance of iodine removal from industrial emissions due to its detrimental effects on human health and the environment. They emphasize the relevance of self-priming venturi scrubbers as a potential solution for efficient iodine abatement, given their ability to create high turbulence and facilitate mass transfer.

Through a series of experimental investigations, the paper explores the hydrodynamic behavior of the self-priming venturi scrubber. The authors discuss the impact of various operating parameters, such as gas flow rate, liquid-to-gas ratio, and throat diameter, on the hydrodynamic characteristics of the scrubber. They present detailed results, including pressure drop, liquid film thickness, and bubble size distribution, offering valuable insights into the fluid dynamics within the scrubber system.

Furthermore, the paper examines the performance of the self-priming venturi scrubber in iodine removal. Ahad et al. [9] discuss the iodine removal efficiency under different operating conditions, highlighting the factors that influence its effectiveness. They also discuss the mechanisms involved in iodine capture, including gas--liquid absorption and chemical reactions, providing a comprehensive understanding of the scrubber's iodine removal capabilities.

12.5.4 Performance Improvement of Venturi Wet Scrubber

Venturi wet scrubbers though they are inexpensive to construct, the operating cost is high due to large pressure drops caused by friction. Kousalya Devi et al. [8] investigated that by increasing the diameter of the scrubber and decreasing the length the pressure drop can be minimized. Further through CFD simulation, the limiting velocity is also identified to improve the scrubber performance. Kousalya Devi et al. [8] implemented Taguchi's L16 orthogonal array for designing the experiment. The authors considered three factors and three levels for finding the pressure drop at various combinations. Through CFD simulation, the flow through the scrubber has been analyzed for finding the limiting velocity. The pattern of contribution for various levels of parameters to the drop in pressure is obtained using the graph obtained.

12.6 FILTRATION FILTERS

Filtration filters are membranes with pores that remove suspended solid matter from a liquid when it is passes through it with force. Filtration process keeps water, pharmaceuticals, and chemicals clean and free from contaminants [10] (Figure 12.5).

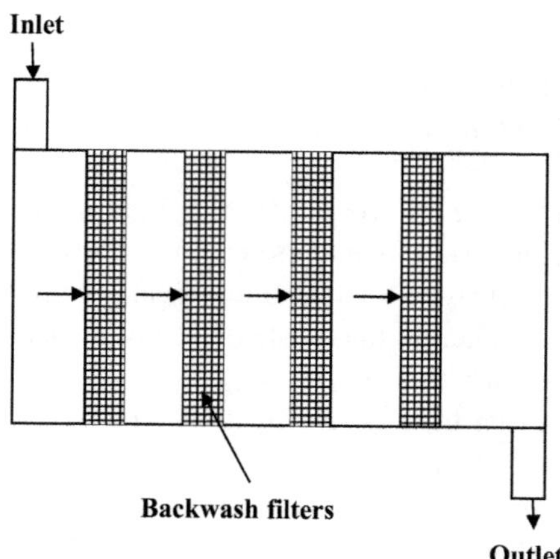

FIGURE 12.5 Backwashing filter during filtration.

12.6.1 Working of Filtration Filters

1. The basic principle of filtration is the mechanical obstruction of particles as the fluid passes through a porous medium. The filter material has pores that allow the passage of the fluid while trapping and retaining particles based on their size.
2. Filters are selected based on their pore size, which determines the size of particles they can effectively capture.
3. In depth filtration, the separation occurs when particles are trapped within the depth of the filter medium.
4. In surface filtration, particles are captured on the surface of the filter medium.
5. Materials including cellulose, fiberglass, polypropylene, ceramic, and metal can be used as filter media in the filtration process.

12.6.2 Roles of Filtration Filters in Industries

1. Water Treatment: Filtration is widely used in water treatment processes to remove suspended particles, sediments, and contaminants from raw water, ensuring that the water meets quality standards for consumption and industrial processes.
2. Oil and Gas Industry: Filtration is utilized in the oil and gas industry to remove solid particles, debris, and contaminants from crude oil, natural gas, and hydraulic fluids to protect equipment and ensure product quality.
3. Automotive Industry: Automotive applications include oil filtration in engines and transmission systems, air filtration for cabin air and engine intake, and fuel filtration to remove particles and contaminants.
4. Air Pollution Control: Filtration is employed in air pollution control systems to remove particulate matter, dust, and pollutants from industrial exhaust gases before they are released into the atmosphere.
5. HVAC Systems: Air filters are used in heating, ventilation, and air conditioning (HVAC) systems to remove dust and other particles from the air, ensuring indoor air quality.

12.6.3 Filter Cloth Deformation during Backwashing Filtration

The description of filter cloth deformation during backwashing filtration is a critical aspect of understanding and optimizing the performance of filtration systems. As an expert in filtration processes, Morsch et al. [10] provided a concise review of this phenomenon.

During backwashing filtration, the filter cloth undergoes deformation because of various factors, including the differential pressure across the cloth, the characteristics of the filtrate, and the filtration cycle. Understanding and characterizing these deformations are crucial for ensuring efficient and effective filtration operations.

Filter cloth deformation primarily occurs due to the accumulation of particulate matter or solids on the cloth surface. As the filtration cycle progresses, the solids build up and form a filter cake on the cloth, leading to increased resistance to flow and pressure differentials across the cloth. This pressure differential causes the cloth to deform, leading to changes in its pore structure and permeability.

The deformation of the filter cloth during backwashing filtration has both positive and negative implications. On one hand, the deformation can help enhance the filtration process by promoting cake formation and improving particle capture efficiency. It can also lead to an increase in the effective filtration area, thus extending the service life of the cloth.

To optimize filtration operations, it is essential to monitor and control filter cloth deformation during backwashing. This can be achieved through proper selection of filter cloth materials, understanding the filtration process parameters, and implementing appropriate cleaning and maintenance practices. Additionally, advancements in filter cloth technology, such as the development of robust and flexible materials, can help mitigate deformation issues and improve filtration performance.

12.6.4 Regeneration Assessments of Filter Fabrics in Mining Sector Filter Presses

The regeneration assessments of filter fabrics in filter presses within the mining sector are of paramount importance for optimizing filtration processes and minimizing operational costs. As an expert in filtration technologies for the mining industry, Frankle et al. [11] provided a concise review of this topic.

Filter presses are widely used in the mining sector for dewatering mineral concentrates and tailings. The filter fabrics play a critical role in this process, serving as the filtration medium for separating solids from liquids. Over time, filter fabrics become clogged with solids and require regeneration to restore their filtration efficiency.

Regeneration assessments involve evaluating the condition and performance of filter fabrics and determining the optimal regeneration methods. These assessments typically include analyzing parameters such as permeability, cake release properties, fabric integrity, and the extent of fouling or blinding.

Efficient regeneration of filter fabrics is crucial for maintaining continuous filtration operations and reducing downtime. Various regeneration techniques are employed in the mining sector, including chemical cleaning, mechanical cleaning, and thermal cleaning. The selection of the most suitable regeneration method depends on factors such as fabric type, extent of fouling, process requirements, and economic considerations.

The assessment of filter fabric regeneration involves a combination of laboratory testing, pilot-scale studies, and field trials. These assessments provide insights into the effectiveness of different regeneration techniques, allowing for the optimization of regeneration protocols and the selection of appropriate cleaning agents or methods.

Optimizing the regeneration process offers several benefits to the mining industry. It extends the service life of filter fabrics, reducing the frequency of fabric replacements and associated costs. Efficient regeneration also ensures consistent filtration performance, enabling the production of high-quality dewatered products and minimizing the environmental impact of mining operations.

Furthermore, advancements in filter fabric materials and coatings have enhanced their regeneration capabilities. Innovative fabrics with enhanced cake release properties and resistance to fouling have resulted in improved regeneration outcomes, reducing downtime and increasing productivity.

12.7 DRAFT FAN

Draft fans are fans used for supplying pressurized air to the system. The induced draft fan removes the smoke and fly ash generated in the combustion chamber through the ESP of steam boilers and thermal oil heaters [12] (Figure 12.6).

12.7.1 Working of Draft Fan

1. The draft fan operates by creating a negative pressure or suction within a system. It pulls air and gases from the combustion chamber, boiler, or other industrial processes, helping to enhance the flow of exhaust gases.
2. Centrifugal draft fans use blades mounted on a rotating impeller to impart kinetic energy to the air or gases. The air is then directed tangentially into a spiral housing, converting the kinetic energy into potential energy and creating a pressure difference.
3. Axial draft fans have blades that generate airflow parallel to the fan's axis. They are suitable for applications where a straight-through airflow is required.
4. Draft fans are often equipped with control systems to regulate the airflow based on the specific requirements of the industrial process. This ensures that the negative pressure is maintained at the desired level for efficient exhaust gas extraction.

12.7.2 Roles of Draft Fan in Industries

1. Boilers and Furnaces: One of the primary applications of draft fans is in boilers and furnaces, where they help in creating the necessary draft for efficient combustion. They extract the combustion gases, including carbon dioxide and other byproducts, and assist in maintaining a stable and controlled combustion process.

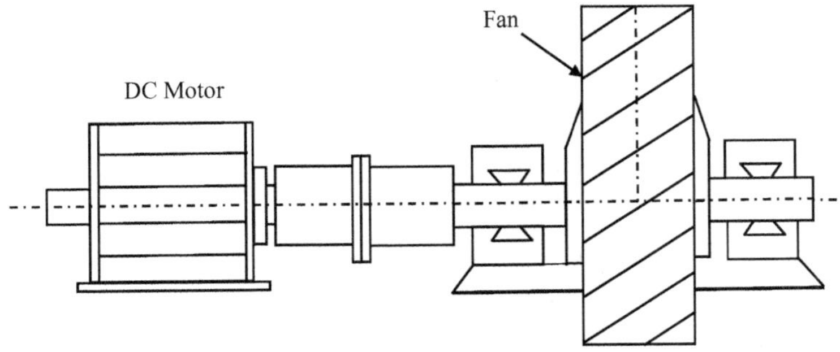

FIGURE 12.6 Schematic diagram of vibration points in an ID fan.

2. Power Plants: Draft fans are integral components in power plants, including thermal power plants that rely on combustion processes for electricity generation. They help in maintaining the required draft in the combustion chambers and boilers.
3. Pulp and Paper Industry: Draft fans assist in extracting exhaust gases from various processes in the pulp and paper industry, such as boilers, digesters, and lime kilns.
4. Cement Industry: Draft fans are used in cement plants to extract exhaust gases from the kiln and other production processes. They contribute to the removal of pollutants and facilitate efficient operation

12.7.3 Vibrational Analysis of the Induced Draft Fan

The research paper titled "Vibration Analysis of Induced Draught Fan: A Case Study" presents an insightful examination of vibration analysis techniques applied to an induced draught fan system. As experts in the field, Dutta et al. [12] contributed to the understanding and diagnosis of fan vibration issues.

The authors begin by providing a concise introduction, outlining the significance of induced draught fans in various industrial applications and their critical role in maintaining optimal system performance. They highlight the importance of vibration analysis as a diagnostic tool for detecting and resolving mechanical faults and abnormalities in fan systems.

The case study presented in this paper showcases the practical application of vibration analysis in assessing the health and performance of an induced draught fan. The authors describe the methodology employed, which includes the use of vibration sensors, data acquisition techniques, and signal processing algorithms to analyze the vibration signatures of the fan components.

The results and findings of the case study are well-documented, with a clear presentation of vibration data and analysis. The authors effectively demonstrate how vibration characteristics can be interpreted to identify specific faults or abnormalities in the fan system. Furthermore, they discuss the implications of these findings in terms of potential performance degradation, maintenance requirements, and overall system reliability.

One strength of this paper is its emphasis on practical implementation and real-world relevance. By providing a detailed case study, the authors offer valuable insights into the challenges and complexities associated with vibration analysis of induced draught fans. This enables readers, particularly maintenance engineers and practitioners, to gain a better understanding of the techniques and considerations involved in diagnosing and mitigating vibration-related issues.

12.7.4 Regeneration Assessments of Filter Fabrics in Mining Sector Filter Presses

The regeneration assessments of filter fabrics in filter presses within the mining sector are of paramount importance for optimizing filtration processes and minimizing operational costs. As an expert in filtration technologies for the mining industry, Frankle et al. [11] provided a concise review of this topic.

Filter presses are widely used in the mining sector for dewatering mineral concentrates and tailings. The filter fabrics play a critical role in this process, serving as the filtration medium for separating solids from liquids. Over time, filter fabrics become clogged with solids and require regeneration to restore their filtration efficiency.

Regeneration assessments involve evaluating the condition and performance of filter fabrics and determining the optimal regeneration methods. These assessments typically include analyzing parameters such as permeability, cake release properties, fabric integrity, and the extent of fouling or blinding.

Efficient regeneration of filter fabrics is crucial for maintaining continuous filtration operations and reducing downtime. Various regeneration techniques are employed in the mining sector, including chemical cleaning, mechanical cleaning, and thermal cleaning. The selection of the most suitable regeneration method depends on factors such as fabric type, extent of fouling, process requirements, and economic considerations.

The assessment of filter fabric regeneration involves a combination of laboratory testing, pilot-scale studies, and field trials. These assessments provide insights into the effectiveness of different regeneration techniques, allowing for the optimization of regeneration protocols and the selection of appropriate cleaning agents or methods.

Optimizing the regeneration process offers several benefits to the mining industry. It extends the service life of filter fabrics, reducing the frequency of fabric replacements and associated costs. Efficient regeneration also ensures consistent filtration performance, enabling the production of high-quality dewatered products and minimizing the environmental impact of mining operations.

Furthermore, advancements in filter fabric materials and coatings have enhanced their regeneration capabilities. Innovative fabrics with enhanced cake release properties and resistance to fouling have resulted in improved regeneration outcomes, reducing downtime and increasing productivity.

12.8 CONCLUSION

Thus, this chapter clearly reveals the functioning and application of the advanced emission control techniques and its development for separating the contaminants from the working medium. Cyclone separator is a device used for the separation of solid particles from a gas or liquid stream based on their inertial forces. Using heterogeneous condensation preconditioning, hydrophilic particles ranging from 0.2 to 2 μm can be separated with 30% improved efficiency with reduced particle size distribution [13]. Moreover, for separating solid particles at elevated temperatures, a convergent vortex finder can be adapted for increasing the tangential velocity, thereby enhancing the centrifugal force correspondingly improving the separation efficiency. An ESP is an emission control device widely used in various industries to remove fine particulate matter, such as dust and ash, from exhaust gases. Introduction of turning vanes and redesign of the electrostatic precipitator with a tube-type arrangement lead to enhancement in flow distribution and minimizes turbulence. The performance of ESP will improve if it is operated using high voltage say 20 kV instead of 12 kV, so that particulate matter of size less than 1 μm can be precipitated. A CO_2 capturing system is a technology that captures carbon dioxide

(CO_2) emissions from various sources such as power plants or industrial processes, preventing them from being released into the atmosphere and contributing to climate change. The optimum parameter for maximum CO_2 removal rate is an ammonia concentration below 14%, a reaction temperature below 20°C, and a reaction time of 30 minutes. Using high-weight pipelines, the gas can be transported over 1,000 km and stored in muddy bowls in land or seaward.

Scrubber system is an air pollution control equipment used to filter the industrial exhaust stream to be free from particulates. The performance of the venturi scrubber can be enhanced by increasing the diameter of the scrubber and decreasing the length, thereby minimizing the pressure drop. Filtration filters are membranes with pores that remove the suspended solid matter from a liquid when it is passing through it with force. In addition, the development of robust and flexible materials can help mitigate deformation issues and improve filtration performance. Draft fans are fans used for supplying pressurized air to the system. Using vibration sensors, data acquisition techniques and signal processing algorithms vibration signatures of the fan components can be analyzed.

REFERENCES

1. J. Wang, X. Duan, S. Wang, J. Wen, J. Tu, "Experimental and numerical investigation on the separation of hydrophilic fine particles using heterogeneous condensation preconditioning technique in gas cyclones", *Separation and Purification Technology*, 2021, 259, 118126.
2. A. Jafarnezhad, H. Salarian, S. Kheradmand, J. Khaleghinia, "Performance improvement of a cyclone separator using different shapes of vortex finder under high-temperature operating condition", *Journal of the Brazilian Society of Mechanical Sciences and Engineering*, 2021, 228, 145–152.
3. E. Jaroudi, I. Sretenovic, G. Evans, H. Tran, "Factors affecting particulate removal efficiency of kraft recovery boiler electrostatic precipitators: a technical review", *TAPPI Journal*, 2018, 17, 273–283.
4. C. Bhasker, "Flow simulation in electro-static-precipitator (ESP) ducts with turning vanes", *Advances in Engineering Software*, 2011, 42, 501–512.
5. A.K.S. Parihar, T. Hammer, G. Sridhar, "Development and testing of tube type wet ESP for the removal of particulate matter tar from producer gas", *Renewable Energy*, 2015, 74, 875–883.
6. A.S. Hamouda, M.S. Eldien, M.F. Abadir, "Carbon dioxide capture by ammonium hydroxide solution and its possible application in cement industry", *Ain Shams Engineering Journal*, 2020, 11, 1061–1067.
7. A.K. Shukla, Z. Ahmad, M. Sharma, G. Dwivedi, T.N. Verma, S. Jain, P.V.A. Zare, "Advances of carbon capture and storage in coal-based power generating units in an Indian context", *MDPI Energies,* 2020, 13, 4124.
8. S. Kousalya Devi, S. Venkatesh, S. Chandrasekaran, "Performance improvement of venturi wet scrubber", *IJRDO-Journal of Mechanical and Civil Engineering*, 2015, 1, 127–135.
9. J. Ahad, T. Rizwan, A. Farooq, K. Waheed, M. Ahmad, K.R. Qureshi, W. Siddique, N. Irfan, "Study of hydrodynamics and iodine removal by self-priming venture scrubber", *Nuclear Engineering and Technology*, 2022, 55, 169–179.
10. P. Morsch, M. Bauer, C. Kessler, H. Anlauf, H. Nirschl, "Description of the filter cloth deformation during backwashing filtration", *Separation and Purification Technology*, 2020, 253, 117504.

11. B. Frankle, P. Morsch, H. Nirschl, "Regeneration assessments of filter fabrics of filter presses in the mining sector", *Minerals Engineering*, 2021, 168, 106922.
12. R. Dutta, J.P. Dwivedi, V.P. Singh, A. Ghosh, "Using vibration analysis to identify & correct an induced draft fan's foundation problem of a pollution control device – A case study", *International Journal of Applied Engineering Research*, 2018, 13, 5831–5840.
13. A. Vimal, S. Thalaieswaran, N.H. Kannan, P. Ganeshan, S. Venkatesh, "A review on the investigation of hydrocyclone performance by shape optimization", *E3S Web of Conferences*, 2023, 405, 04047.

13 Energy Efficiency Policies, Regulations, and Financing in the Industry Sector

Manjushree Banerjee and Bhaskar Natarajan

13.1 INTRODUCTION

Energy efficiency is the first step that results in the reduction of the demand for energy and is the first strategy in an effort to solve the issues posed by pollution and climate change. The direct benefits of energy efficiency are cost savings on fuel expenditure and reduced pollution at the environmental level. Indirectly, energy efficiency in the industries creates a more resilient and reliable energy security scenario at the macro level and overall health benefits for the broader community. There are negotiations at international levels for climate justice, where energy efficiency plays a crucial role. At the country level, there are policies, regulations, and financing mechanisms to advance the energy efficiency pathways. Energy efficiency measures are considered across the entire range of consumers in the transport, building, appliances, and industrial sectors.

At the global level, the industry sector consumes about 170 EJ, which is one-third of the total final energy consumption (IEA, 2023). Globally, coal accounts for about 28% (IEA, 2023) of energy consumed by the industrial sector. According to this, electricity is responsible for supplying 23% of the energy that is used in the industrial sector, and this percentage is growing. Approximately 37% of the energy that is used in the industrial sector comes from natural gas and oil. Steel, cement, and aluminum are the three industries that absorb 60% of the total energy that is consumed by the industrial sector.

In order to attain net-zero emissions by the year 2070, India is introducing renewable energy. The industrial sector consumed about 56% of energy in the year 2020–21 (Bureau of Energy Efficiency, 2022). The final energy consumption of India was 553,971 ktoe. As a part of climate mitigation policies, the government of India has set targets for the installation of RE-generation capacities, as well as plans for shaping an energy-efficient economy. In assessments of energy efficiency programs, the Bureau of Energy Efficiency (BEE) mentions a significant decline in energy intensity from 0.27 MJ per INR in the year 2011–12 to 0.22 MJ per INR in the year 2020–21 (Bureau of Energy Efficiency, 2022). In 2020–21, the energy efficiency schemes/programs resulted in an aggregate energy savings of 42.0 Mtoe, which is equivalent to 4.73% of the primary energy.

DOI: 10.1201/9781003452072-13

Industrial energy efficiency measures across the world follow some major options, such as using energy-efficient technologies for motors and drive systems, recycling materials, material substitutions, switching to renewable fuels, and planned energy management systems. The present chapter discusses the policy and regulatory environment, especially in India, to enhance energy efficiency in industries. Furthermore, the chapter looks at various technical and financial mechanisms enabling the transition to greener fuels.

13.2 GLOBAL SCENARIO

Industrial energy consumption at the global level accelerated at a yearly pace of 1.3% from the year 2010 to 2022 (IEA, 2023). China consumes 30% of the total energy consumption, while the EU, India, Japan, and the United States account for about 28% of the total industrial energy (Figure 13.1).

With the advent of the energy crisis in the year 2022, investments in energy efficiency measures increased at the global level. As per the IEA's Government Energy Spending Tracker, about 70% of the total investments in energy efficiency, which is about USD 700 billion, in the year 2020, were made by five countries (IEA, 2023). The United States included USD 86 billion in energy efficiency measures Inflation Reduction Act of 2022 (IEA, 2023).

While pursuing energy efficiency policies, China set targets to enhance energy intensity by 13.5% by 2025 compared to the 2020 baseline. The targets include 17 energy-intensive industries and options such as upgrading electric motors and transformers.

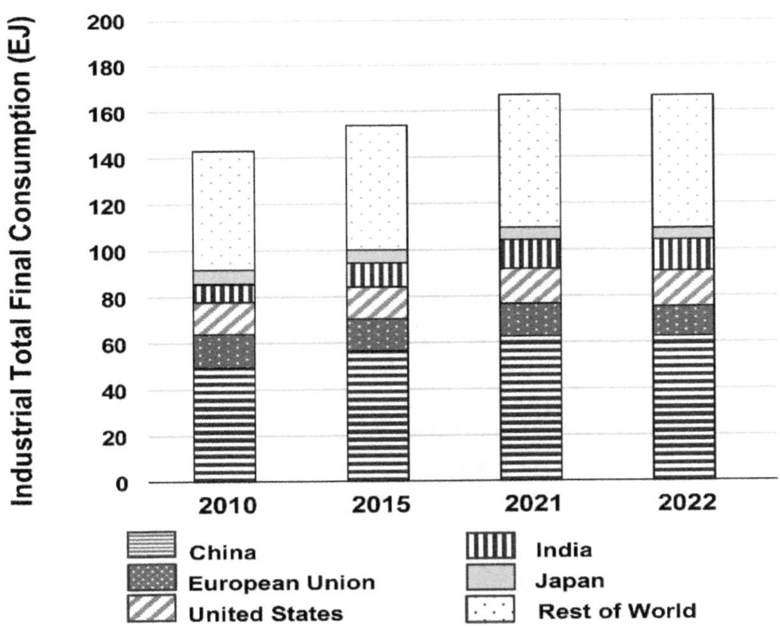

FIGURE 13.1 Total final energy consumption for industry, 2010–2022. (IEA, 2023.)

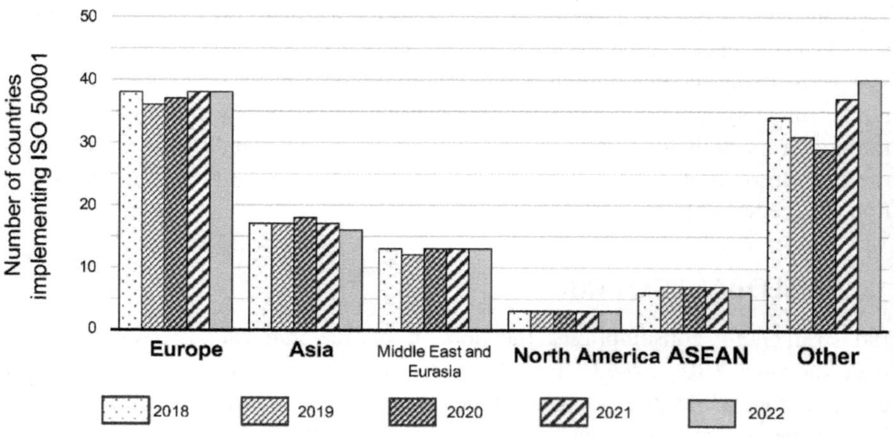

FIGURE 13.2 ISO 50001 certifications by region, 2018–2022. (IEA, 2023.)

Globally, light industries (such as food, construction, and machinery) consume more electricity-based energy compared to heavy industries. In the year 2023, about 62 countries adhered to Minimum Energy Performance Standards (MEPS) for enhancing motor efficiency in their industries (IEA, 2023). The International Electrotechnical Commission (IEC) sets international standards for electrical, electronic, and related technologies worldwide (IEC, 2024).

The International Organization for Standardization (ISO) gathers data on the certificates issued. As per the IEA, the ISO 50001 is among the more effective tools for tracking energy savings in industries. In the year 2022, about 28,000 ISO 50001 certificates were issued. Figure 13.2 provides the distribution of countries/regions based on their issuance of ISO 50001 certificates.

The Government of Ras Al Khaimah became the first country in the world in the year 2023 to issue ISO 50001 certificates to all the entities in its country (IEA, 2023). The country reported a 25% savings of electricity during the period (IEA, 2023). Carbon pricing is also followed by many countries to tax the burden of emissions on those considered to be responsible. In the case of ETS, the low-emission-producing industries may sell their certificates to industries with high emission levels. Indian Energy Exchange (discussed later in this chapter) is one of the examples of ETS. Carbon taxes define the price of carbon in the form of tax. Europe's Carbon Border Adjustment Mechanism levies carbon tax on a few imported goods such as steel to ensure that the products they import meet certain emission norms. As per the Carbon Pricing Dashboard, in the year 2023, these initiatives were expected to cover about 11.66 GtCO2e, i.e., 23% of the world's GHG emissions.

In the present world's context, carbon pricing, MEPS, public procurement of low-emissions materials, and funds for Industrial Innovations form the major categories of the policies, regulations, and investments toward energy efficiency in industries. Table 13.1 indicates the implementation of such initiatives for industrial energy efficiency by country.

TABLE 13.1
EE Policies and Regulations for Industries

	Carbon pricing	Minimum energy performance standards for motors	Public procurement of low-emissions materials	Funds for industrial innovation
United States	100%	100%	100%	100%
Latin America and the Caribbean	<10%	10-39%	<10%	10-39%
European Union	100%	70-99%	70-99%	100%
Africa	<10%	<10%	<10%	10-39%
Middle East	<10%	10-39%	<10%	<10%
Eurasia	<10%	<10%	<10%	10-39%
China	100%	100%	100%	100%
India	<10%	100%	<10%	<10%
Japan and Korea	100%	100%	100%	100%
Southeast Asia	<10%	10-39%	<10%	<10%

Share of countries within each region with policy currently implemented:
<10% 10-39% 40-69% 70-99% 100%

Source: IEA (2023).

13.3 INTERNATIONAL COOPERATION

The G20 Energy Transition Working Group, presided over by India in the year 2023, identified six priority areas (Ministry of Power, 2023). The Conference of Parties (COP) 28, organized at the year-end 2023, hosted dialogues on accelerating green procurement commitments and financing mechanisms toward decarbonization. The Working Party on Industrial Decarbonisation has its focus on definitions and measurement methodologies for heavy industries. These definitions and measurements build upon the G7 presidencies of 2022 and 2023 (IEA, 2023).

13.4 THE CONTEXT OF INDIA

From the year 2010 to 2022, industrial energy consumption in India grew by 4.5% annually, which is significantly higher than the global rate of 1.3% (IEA, 2023).

There are two important ministries in the areas relating to emissions control and energy efficiency. First, the Ministry of Environment, Forests & Climate Change (MOEF&CC) is responsible for all issues pertaining to the impact of economic activity on the environment, including solid, liquid, and gaseous emissions. MoEF&CC operates majorly through the Central Pollution Control Board (CPCB) and the State

Pollution Control Boards (SPCB) for controlling all emissions from the industries. The CPCB and SPCBs are the statutory organizations enforcing compliance with the various laws in connection with discharges into the environment from any economic activity. Second, the Ministry of Power coordinates all matters relating to energy efficiency and conservation in India. The BEE was set up with powers from the Energy Conservation Act, 2001.

On the judicial side, while the Indian courts exist to pronounce verdicts on any issues relating to any legislation in the country, the National Green Tribunal has been established with branches across India to adjudicate on all matters specifically coming under any of the legislations related to the subject matter of environment. Also, matters relating to energy efficiency are adjudicated by the Appellate Tribunal for Electricity.

The environmental legislations prescribe rules for the discharges into the environment, to be followed by industries, buildings, offices, hotels, and other organizations. As against this, the BEE prescribes targets for the industries and other designated consumers to adopt and implement strategies that optimize energy consumption.

13.4.1 Energy Conservation Act 2001

Clause 14 of the Energy Conservation Act 2001 requires the government to set energy consumption standards for specified users to encourage energy efficiency. These rules apply to certain energy-intensive sectors and establishments listed in the Act. The law sets energy consumption benchmarks to decrease energy use and promote sustainability across industries. This regulatory framework is essential for energy conservation and resource efficiency in industry and commerce. The Act also specifies the intensity of energy usage and investments needed to switch to energy-efficient equipment. The BEE was established in 2002 in accordance with the Energy Conservation Act of 2001. The objective of BEE is to design methods for the self-regulation and dynamics of the market. The main goal is to achieve energy efficiency, leading to a decrease in the country's energy usage.

13.4.2 National Action Plan for Climate Change

The National Action Plan for Climate Change (NAPCC) was established in 2008 with the objective of achieving sustained economic growth while simultaneously increasing the quality of life for most of the population and minimizing their susceptibility to the implications of climate change (Government of India, 2008). The National Action Plan comprised eight distinct missions, namely: National Solar Mission, National Mission for Enhanced Energy Efficiency, National Mission for Sustainable Habitat, National Water Mission, National Mission for Sustaining the Himalayan Ecosystem, National Mission for Green India, National Mission for Sustainable Agriculture, and National Mission for Strategic Knowledge for Climate Change. The Ministry of Power and (BEE developed the implementation strategy and framework for the National Mission for Enhanced Energy Efficiency (NMEEE).

13.4.3 National Mission for Enhanced Energy Efficiency (NMEEE)

The NMEEE aims to transition to cleaner fuels and transfer commercially feasible efficient technologies following market-based mechanisms. Energy Efficiency Services Ltd. (EESL) was formed in the year 2009 as an implementing arm for EE that will have close coordination with BEE and works under the overall administrative and policy guidance of the Ministry of Power (MoP) and BEE.

NMEEE comprises four efforts aimed at augmenting energy efficiency in businesses with high energy consumption: (1) The Performance Assessment Tool aims to enhance energy efficiency in sectors that consume a lot of energy. (2) The Energy Efficiency Financing Program (EEFP) helps stakeholders involved in energy efficiency financing to improve their skills and knowledge. (3) The Framework for Energy Efficient Economic Development (FEEED) focuses on creating financial tools to encourage energy efficiency. (4) The Market Transformation for Energy Efficiency (MTEE) program aims to speed up the adoption of energy-efficient appliances (Figure 13.3).

13.4.4 Perform, Achieve, and Trade (PAT)

About 25% of the total energy in India is covered by Perform, Achieve, and Trade (PAT) Scheme (International Energy Agency, 2021). The PAT plan aims to decrease energy

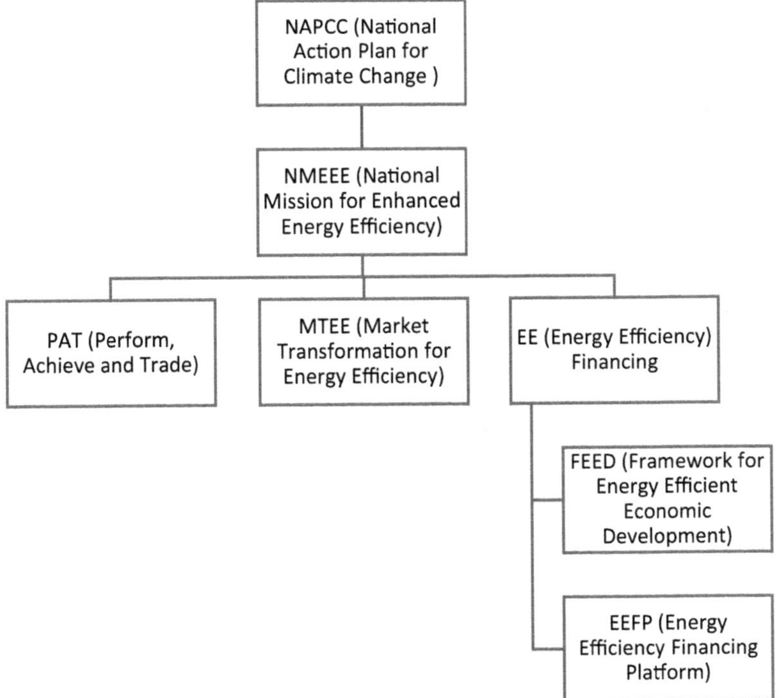

FIGURE 13.3 Components of NMEEE. (Author based on information compiled in the chapter.)

TABLE 13.2
Energy, Emission and Money Saved during the PAT Cycles

PAT Cycle	Electricity Saving (BU)	Thermal Saving (Mtoe)	Total Energy Savings (Mtoe)	GHG Reduction ($MtCO_2$)	Money Savings (INR crore)
III	0.61	1.69	1.75	6.43	3,483
II	36.47	10.95	14.08	68.64	42,020
I	3.01	8.41	8.67	31.00	9,500

Source: Bureau of Energy Efficiency (2022).

usage in the designated energy-intensive industries, referred to as designated consumers (DCs). The scheme identified 20 large energy-intensive sectors following the NMEEE and Energy Conservation Act 2001. The Government decided on the targets for reducing specific energy consumption after following a detailed process of getting information on the present energy consumption, the technology and processes for manufacturing, and the potential for saving energy. The scheme has multi-cycles and each cycle is of 3 years. The Energy Savings Certificates (ESCerts), equivalent to 1 ton of oil equivalent (toe) of energy savings, are awarded post-verification and can be sold to another DC or may be banked for the next PAT cycle. There are penalties linked to non-compliance. Several organizations are involved in the implementation of the plan, including the Central Electricity Regulatory Commission, which is in charge of regulating the market, the Power System Operation Corporation, which is responsible for registering the ESCerts, and the BEE, which is in charge of managing the trading process.

The PAT was rolled in a series of cycles. The designated consumers list reached 1073 by the sixth cycle (Bureau of Energy Efficiency, 2022). The savings in terms of energy, emissions, and money are presented in Table 13.2.

Case Study (Steel Industries)

In the iron and steel industry, the Energy Conservation Act of 2001 limits unit energy use to 20,000 tons of oil equivalent (toe) per year. This regulation aims to improve energy efficiency and minimize energy consumption in one of the world's most energy-intensive sectors. The Perform, Achieve, and Trade (PAT) method sets gate-to-gate (GtG) targets throughout the whole production process, from raw materials to completed goods. This provides efficiency increases throughout iron and steel production, including mining, processing, refining, and manufacturing. The Indian steel industry was successful in attaining these energy consumption objectives during PAT Cycles 1 and 2. This achievement shows the sector's dedication to innovative technology, process optimization, and energy efficiency. By exceeding these standards, the industry decreases its carbon footprint, lowers operational costs, and improves sustainability, making it more competitive. Energy efficiency targets in the iron and steel sector support India's sustainable industrial growth and climate mitigation aims. It shows how governmental measures and business cooperation may enhance energy management and environmental stewardship nationwide (Ministry of Steel, 2023).

According to the Ministry of Steel, the Indian Steel Industry achieved a 0.5 Ton/ton of crude steel (T/tcs) reduction in energy consumption and carbon emissions from 2005 to 2.6 T/tcs by 2020. The decrease in emissions is a result of implementing state-of-the-art technology and updating the projects to meet contemporary standards (Ministry of Steel, 2022).

Source: Ministry of Steel (2023), Ministry of Steel (2022)

Trading on ESCerts started at the Indian Energy Exchange in 2017 (Figure 13.4). During the period from 2017 to 2018, 86.11 lakh ESCerts were offered for sale

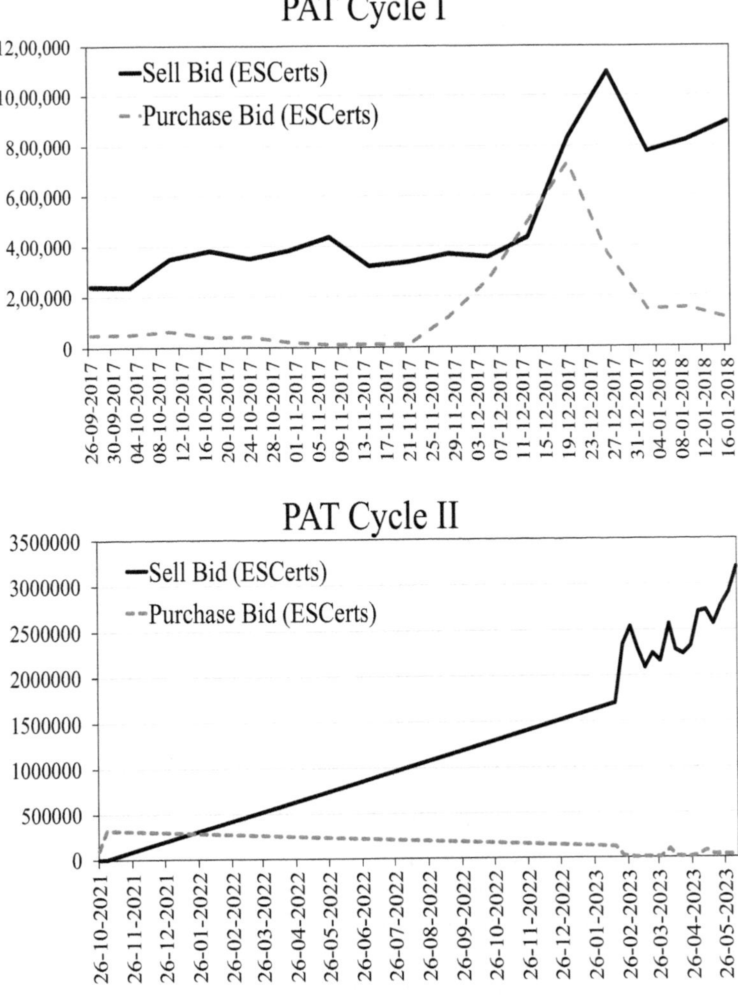

FIGURE 13.4 Sell bids and purchase bids of ECerts. (Author based on the data at Indian Energy Exchange, 2023.)

(Natarajan, Kumar, Rajah, & Koshy, 2021). The ESCerts trading for PAT Cycle I was initiated in January 2017 and closed in January 2018. Trading of ESCerts for PAT Cycle II was initiated in October 2021 (Indian Energy Exchange, 2023). The figure clearly indicates that the sell bids are much higher than the purchase bid, and the gap between the two is increasing in the PAT II Cycle.

The PAT program will continue to be expanded to other energy-intensive sectors and also will be scaled to include more industries and designated consumers at lower levels of target energy consumption, thus bringing more designated consumers into the PAT program. A large part of the savings in the PAT program comes from boilers, utilities, and heating and drying systems, which also contribute to a large share of the emissions from any manufacturing unit.

It is important to note that there are a number of equipment available, such as air pre-heaters, economizers, additional insulation for steam distribution, furnaces, and drying systems, all of which improve the efficiency of the operations, leading to reduced energy consumption and thus reduced emissions.

These are standard technologies that have been available for several years now. Industries prefer to cut costs and discharge into the environment instead of using these technologies to reduce emissions. There is a need for incentive programs to allow industries to install emission reduction technologies instead of controlling the emissions at the tail-pipe which is always more expensive and draws the attention of the pollution boards.

13.4.4.1 Framework for Energy-Efficient Economic Development

FEEED is designed to ease the financing of energy-efficient projects through various fiscal means. Schemes such as the Partial Risk Guarantee Fund (PRGF) address credit-related risks and capacitate financial institutions in financing commercially viable EE projects. The Energy Service Companies (ESCO) and other companies that plan to initiate EE projects are the potential beneficiaries. There are initiatives for Central Public Sector Undertakings to take up energy efficiency initiatives. The FEEED also emphasizes the verification, monitoring, and evaluation of demand-side management aspects for Electricity Regulatory Commissions.

13.4.4.2 Energy Efficiency Financing Platform

BEE has set up an online Center for encouraging and upscaling energy efficiency financing in the country. The objective of the Center is to spread awareness on financing (BEE, 2023). The BEE website lists about 179 energy-efficient technologies. The EEFP aims to interact with institutions for energy efficiency projects. Over 300 companies have signed up their willingness to invest in energy efficiency.

13.4.4.3 Market Transformation for Energy Efficiency (MTEE)

The MoP (MTEE) promotes energy-efficient appliances. Bachat Lamp Yojana (BLY) replaced incandescent lights with compact fluorescent lamps (CFLs) priced equally. CFLs are energy efficient and last longer than incandescent lamps, therefore this program promoted their use.

MTEE's Super-Efficient Equipment Program (SEEP) aims to revolutionize high-efficiency appliance sales. SEEP promotes energy-efficient appliances by targeting them. SEEP promotes super-efficient appliances to minimize residential and industrial energy consumption and carbon emissions.

These activities demonstrate India's dedication to energy saving across industries. The government encourages energy-efficient technology like CFLs and super-efficient appliances through BLY and SEEP to minimize energy consumption, greenhouse gas emissions, and promote sustainable energy development.

13.4.4.4 Energy Efficiency in MSME Sector

Micro, small, and medium enterprises (MSME) contribute about 30% to the country's gross domestic product (GDP). The sector produces a wide range of goods and thus the technologies and energy consumption levels also spread over a wide range. Although BEE implemented energy efficiency measures in 29 out of 35 clusters, the measures are yet to be mainstreamed. BEE launched the "National Program on Energy Efficiency and Technology Upgradation in SMEs," which promotes the quick adoption of energy-efficient technology and innovative funding mechanisms in India's SME sector to improve energy efficiency. BEE, Global Environment Facility (GEF), and EESL collaborated on this program. Over 75 MSMEs have had energy audits as part of this initiative. These audits are essential for finding energy efficiency improvements and suggesting technology. MSMEs may drastically cut energy usage, operating expenses, and environmental footprint by applying audit findings. The scheme also educates MSMEs about energy efficiency and helps them get funding for technological improvements. This comprehensive strategy boosts MSMEs' competitiveness and supports India's sustainable development and energy security goals. The National Program on Energy Efficiency and Technology Upgradation in SMEs aims to turn India's SME sector into a more energy-efficient and ecologically sustainable one through these collaborations. The dashboard of the program, Promoting Market Transformation for Energy Efficiency in Micro, Small, and Medium Enterprises (MSME) (an initiative of United Nations Industrial Development Organization [UNIDO], Ministry of MSME, GEF, and EESL), shows that to date, 101 EE equipment are installed, which leads to energy saving of 1,986 toe per annum and 9,205 tCO_2 per annum of emission reduction.

13.5 CENTRAL POLLUTION CONTROL BOARD (CPCB)

The Government of India sets standards to control the pollution emitted from industrial sources. Standards are based on the pollutants from specific sources. The National Ambient Air Quality Standards has notified 12 pollutants and 115 emission/effluent standards for 104 different sectors of industries. There are general 32 standards for ambient air CPCB has Standards for the Emission and Discharge for more than 100 Industries. The General Standards for Industries in terms of Particulate Matter (PM) is limited to 150 mg/Nm^3.

India is among the leading countries that have initiated policies for reducing emissions and thus emissions intensity at the national level as a part of voluntary communications with the UN. Emissions control is necessary for both air and liquid emissions. These are controlled by legislations that specify how much emissions can be released by different industries. CPCB, the nodal agency also promotes emissions reductions through the adoption of cleaner technologies in the production processes which would lead to a reduction in both the discharge of gaseous and liquid emissions into the environment.

Industries have largely been complying due to the fear of penalties, fines, and the possibility of a closure notice. The CPCB and SPCBs have been quite diligent in their efforts to control emissions. They have also adopted online systems to monitor online discharges and notices are issued on the basis of the readings and analysis from the online systems.

13.6 TRANSITION TO GREENER FUELS AND TECHNOLOGIES

The transport sector consumes a major share of the total energy. Electrification in the transport sector has remained the thrust in recent years. The sales of electric cars reached almost 80,000 in India, Thailand, and Indonesia in the year 2022. There are many supportive policies to promote electric vehicles (EVs), such as interest-free loans to lower-income households to buy EV vehicles in France. Figure 13.5 indicates the cost and sale of energy efficiency and electrification in the transport sector.
India is focusing on solar photo voltaic (PV) on building out the grid and enhancing the domestic supply chain.

Green Hydrogen is considered a strong alternative for industries to move toward low-carbon growth. The National Green Hydrogen Mission was approved in January 2022. Hydrogen has a vital role in the diversification of energy sectors.

The Petroleum and Natural Gas Regulatory Board (PNGRB) has designated a 5% volume/volume blending ratio of green hydrogen with PNG.

In addition, there are several technologies available that can substitute coal or gas for heating and drying purposes. These are available across a range of sizes for application in a large number of industries such as textiles, paper, leather, and other such industries. There is a need for these technologies to be taken up in pilot mode for demonstration. These technologies will directly contribute to a significant reduction in the emissions of the industries that use coal, oil, and gas to a large extent. It is important for schemes to be put in place for such a conversion to happen in an accelerated manner across the country.

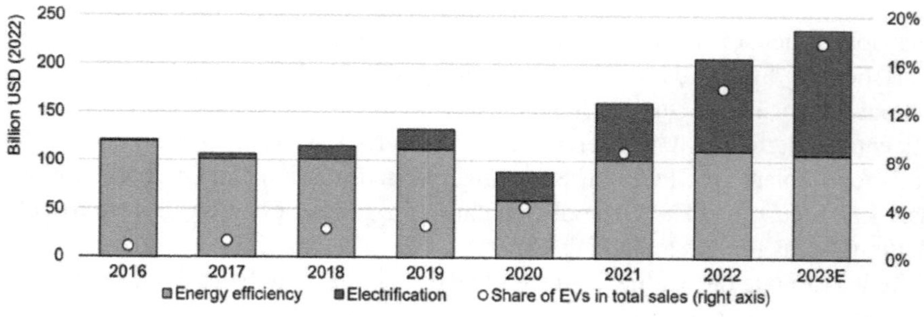

FIGURE 13.5 Energy efficiency and electrification spending in the transport sector. (IEA, 2023.)

13.7 FINANCING

The total investment in energy efficiency globally in 2023 was about USD 620 billion (IEA, 2023). Figure 13.6 indicates the share of overall investments across major sectors. As per IEA, electric vehicles are the major drivers toward an increase in energy efficiency in the year 2023. IEA estimates the need for a triple-fold increase in investments from the current level by 2030 to reach the net-zero emission targets by 2050.

The government of India's first introduced green bond was four times oversubscribed. The fund supported projects aiming at reducing energy intensity and renewable power projects. One advantage of these instruments is accessing the funding sources. A significant number of green bond issuances in India have historically been denominated in US dollars, and it is probable that the government aims to foster the growth of a domestic market through sovereign issuances. The majority of investors were from the local market, while overseas investors appeared to be discouraged by the potential risk associated with currency fluctuations. *Source*: International Energy Agency, 2023.

In April this year, the Reserve Bank of India (RBI) floated a framework for green deposits, which would apply to all scheduled commercial banks. From time to time, different governments have promoted energy efficiency through subsidies. The Japanese Government provides subsidies to the industries for replacing existing equipment with custom-made energy-efficient equipment (IEA, 2022).

13.8 CHALLENGES

There are challenges associated with implementation of the energy efficiency measures in the case of the industry sector. The Energy Systems Tracking Report of

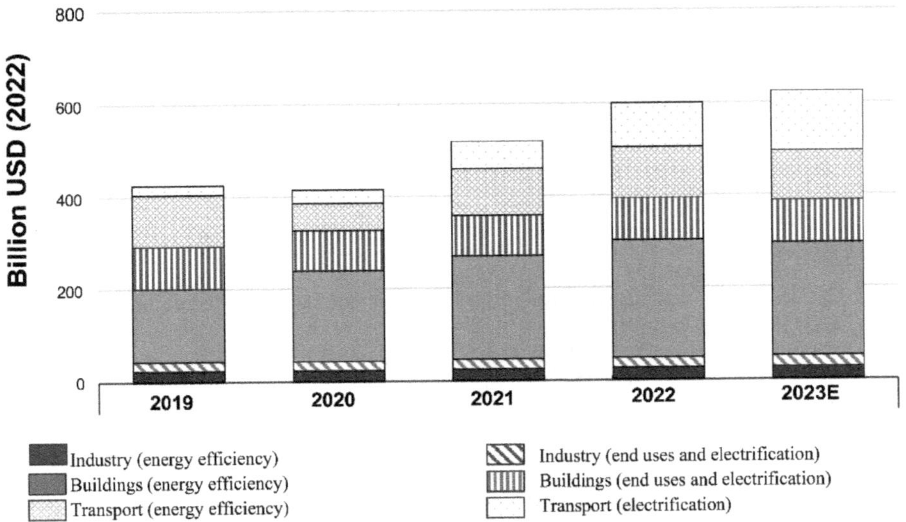

FIGURE 13.6 Global investment in energy efficiency, electrification, and renewables for end uses, 2019–2023(E). (IEA, 2023.)

IEA indicates 'Not on Track' in the case of Industries (IEA, 2023). The industry sector is responsible for emitting 9.0 Gt of CO_2 and the decline in emissions is not adequate to achieve the net-zero emission targets at the global level. One of the major challenges is the long life of the machinery in the industrial sector. Replacing the existing machinery and motors with energy-efficient technologies demands considerable investments and efforts. The Energy Tracking Report 2023 of IEA indicates considering strong government policies to get on track. The report also emphasizes accelerating energy and material efficiency regulations to enhance energy efficiency in the sector.

The Energy System Tracking Report 2023 (IEA, 2023) provides seven sets of recommendations for the industry sector. The report emphasizes taking up mandatory industrial energy transition plans and emission policies such as carbon pricing. At a global level, the report mentions the prevention of carbon leakage through carbon border adjustments as one of the solutions. However, national policies and regulations along with financing mechanisms are the primary solutions for energy efficiency in the industry sector.

13.9 CONCLUSION

Industries being among the major consumers of energy have the potential to significantly contribute toward net-zero emission targets and energy efficiency targets. However, there is no one-fit way. Various policies and regulations along with adequate implementation backed by supportive market and international relations frame the overall energy efficiency scenario for any country.

The policies, obligations, and regulations supported by financing mechanisms have enhanced energy efficiency in the industries. Few countries, especially the major energy consumers, have exhibited strong measures toward energy efficiency. Considering the net-zero emission targets and increasing energy demand estimations, it is important to move ahead with energy efficiency measures, especially in the case of industries. Industries across the world have their unique challenges in adopting energy efficiency measures and regulations, primarily because of the associated financial implications. With time, the policies and regulations also need to adapt to the changing markets and the overall social, economic, and geopolitical scenario.

REFERENCES

BEE. (2023). *BEE's Facilitation Centre*. Retrieved from https://www.adeetie.beeindia.gov.in/bee-facilitation-centre

Bureau of Energy Efficiency. (2022). *Impact of Energy Efficiency Measures for the Year 2020–21*. Retrieved May 25, 2023, from https://beeindia.gov.in/sites/default/files/publications/files/Impact%20Assessment%202020-21_FINAL.pdf

Central Pollution Control Board. (2019). *Industry Specific Standards*. Retrieved from https://cpcb.nic.in/

Government of India. (2008). *National Mission for Enhanced Energy Efficiency*. Retrieved from https://www.nicra-icar.in/nicrarevised/images/Mission%20Documents/National%20Mission%20for%20Enhanced%20Energy.pdf

IEA. (2022). *Percentage of Energy Use Covered by Mandatory Energy Efficiency Policies in India, 2010–2018,*. Retrieved from https://www.iea.org/data-and-statistics/charts/percentage-of-energy-use-covered-by-mandatory-energy-efficiency-policies-in-india-2010-2018

IEA. (2022). *Subsidies for Industry and Commercial Energy Efficiency Investments.* Retrieved from IEA50: https://www.iea.org/policies/13825-subsidies-for-industry-and-commercial-energy-efficiency-investments

IEA. (2023). *Energy Efficiency.* Retrieved from iea.org: https://www.iea.org/topics/energy-efficiency

IEA. (2023). *Energy Efficiency 2023.* IEA. Paris: IEA. Retrieved Feb 16, 2024, from https://iea.blob.core.windows.net/assets/dfd9134f-12eb-4045-9789-9d6ab8d9fbf4/EnergyEfficiency2023.pdf

IEA. (2023). *Executive Summary, Energy Efficiency 2023.* Retrieved from https://www.iea.org/: https://www.iea.org/reports/energy-efficiency-2023/executive-summary

IEA. (2023). *Tracking Industry.* Retrieved from IEA: https://www.iea.org/energy-system/industry#tracking

IEA. (2023). *World Energy Outlook 2023.* Retrieved from www.iea.org

IEC. (2024). *IEC Webstore.* Retrieved from https://webstore.iec.ch: https://webstore.iec.ch/home

Indian Energy Exchange. (2023). *Energy Saving Certificates (ESCerts).* Retrieved from https://www.iexindia.com/marketdata/ESCerts_Market.aspx

International Energy Agency. (2021). *E4 Country Profile: Energy Efficiency in India.* Retrieved from https://www.iea.org/articles/e4-country-profile-energy-efficiency-in-india

International Energy Agency. (2021). *Perform, Achieve, Trade (PAT) Scheme.* Retrieved from iea.org: https://www.iea.org/policies/1780-perform-achieve-trade-pat-scheme

International Energy Agency. (2022). *Energy Efficiency 2022.* IEA. Retrieved from https://iea.blob.core.windows.net/assets/7741739e-8e7f-4afa-a77f-49dadd51cb52/EnergyEfficiency2022.pdf

International Energy Agency. (2022). *Savings from Energy Efficiency in India, 2014–2018.* CC BY 4.0

International Energy Agency. (2023). *World Energy Investment 2023.* IEA Publications. Retrieved May 2023, from https://www.iea.org/: https://iea.blob.core.windows.net/assets/54a781e5-05ab-4d43-bb7f-752c27495680/WorldEnergyInvestment2023.pdf

Ministry of Power. (2023). *NTPC Starts India's First Green Hydrogen Blending Operation in PNG Network.* Retrieved from https://pib.gov.in/PressReleaseIframePage.aspx?PRID=1888334#:~:text=NTPC%20Ltd%20commissions%20India's%20first,Gujarat%20Gas%20Limited%20(GGL)

Ministry of Power. (2023). *Press Release.* Retrieved from Press Information Bureau, GOI: https://pib.gov.in/PressReleaseIframePage.aspx?PRID=1939559

Ministry of Steel. (2022). *Indian Steel Industry Reduces Its Energy Consumption and Carbon Emissions Substantially with Adoption of Best Available Technologies in Modernisation & Expansions Projects.* Retrieved from https://pib.gov.in/: https://pib.gov.in/PressReleaseIframePage.aspx?PRID=1794782

Ministry of Steel. (2023). *Energy and Environment Management in Steel Sector.* Retrieved from https://steel.gov.in/en/energy-environment-management-steel-sector

Natarajan, B., Kumar, S., Rajah, V., & Koshy, C. (2021). *State of Energy Efficiency in India- A compilation of Policies, Priorities and Potential.* New Delhi: Alliance for an Energy Efficient Economy.

Index

For Product Safety Concerns and Information please contact our EU representative GPSR@taylorandfrancis.com Taylor & Francis Verlag GmbH, Kaufingerstraße 24, 80331 München, Germany

Batch number: 10397790

Printed by Printforce, the Netherlands